MAGNETO-RESISTANCE FOR CRYSTALS OF GALLIUM

N.V. VAN DE GARDE & CO'S DRUKKERIJ, ZALTBOMMEL

MAGNETO-RESISTANCE
FOR
CRYSTALS OF GALLIUM

BY

Dr J. W. BLOM

Springer-Science+Business Media, B.V

1950

ISBN 978-94-017-5747-8   ISBN 978-94-017-6119-2 (eBook)
DOI 10.1007/978-94-017-6119-2

CONTENTS

CHAPTER I

GALLIUM, PROPERTIES AND TREATMENT; ADJUSTING AND
MEASURING METHOD

CHAPTER II

RESULTS OF THE MEASUREMENTS WITH THE P_1-CRYSTAL, WHEN
THE FIELD IS PERPENDICULAR TO THE CURRENT

CHAPTER III

RESULTS OF THE MEASUREMENTS WITH P_2-, P_3- AND $P_{2,3}$-
CRYSTALS, WHEN THE FIELD IS PERPENDICULAR TO THE
CURRENT

CHAPTER IV

RESULTS OF THE MEASUREMENTS WITH P_1-CRYSTALS, WHEN THE
ANGLE BETWEEN THE FIELD AND THE CURRENT IS VARIED

CHAPTER V

RESULTS OF THE MEASUREMENTS WITH P_2- AND P_3-CRYSTALS,
WHEN THE ANGLE BETWEEN THE FIELD AND THE CURRENT IS
VARIED

CHAPTER VI

THE KOHLER DIAGRAM FOR GALLIUM

CHAPTER VII

THE FOURIER ANALYSIS OF THE ROTATIONAL DIAGRAMS. THE
FOURIER COMPONENTS IN THE KOHLER DIAGRAM

INTRODUCTION

Following up similar measurements with bismuth by L. S c h u b-
n i k o w, I have studied in the years 1933–1937 under the super-
vision of Prof. W. J. d e H a a s the influence of a magnetic field on
the electric resistance of single-crystals of gallium at low temperatures. For the greater part the results have been published; in
chapters II, III and IV a general review is given. For the rest a more
detailed discussion is given in chapter V.

Since these papers were published K o h l e r has introduced a
new method for plotting the experimental results, known as the
"Kohler diagram". In a double logarithmic scale $\Delta R/R_{0T}$ ($\Delta R =$ the
magnetic increase of the resistance, $R_{0T} =$ the resistance without a
field at the temperature of the measurement) is plotted against
$H(R_{0T}/R_{0°c})$ ($H =$ the field strength, $R_{0°c} =$ the resistance without
a field at 0°C). For most metals a "characteristic function" is then
found which is independent of the crystal which has been used, of the
impurity of the material, and of the temperature.

On suggestion of Prof. C. J. G o r t e r the experimental data
found with gallium have been plotted in Kohler diagrams. The discussion of the results is given in chapter VI.

Following up his former proposal K o h l e r suggested that
$\Delta R/R_{0T}$ should be plotted as a function of $H/(R_{0T}/R_{0\Theta})$ ($R_{0\Theta}$ being the
resistance without a field at the Debije temperature of the metal),
which is referred to as a "reduced Kohler diagram". This is based on
the idea that zero degree centigrade has no special meaning for most
of the physical phenomena, whereas at the Debije temperatures different metals are in corresponding states.

Comparing the characteristic functions found for the different
metals in this way, J u s t i and his collaborators stated that several
metal types may be distinguished, each type occupying a relatively
narrow strip in the reduced Kohler diagram.

Though there are some difficulties in the determination of the

Debije temperature in the case of gallium, the results have also been plotted in a reduced Kohler diagram. The discussion of the comparison with the other metals is found in chapter VI.

We have given special attention to the influence of the orientation of the field with respect to the crystal axes on the magneto-resistance effect.

Considerations based on general symmetry rules, valid for single-crystals, allow us to predict some simplifications which are found when the rotational diagrams (giving ΔR as a function of the angle α between the field and one of the crystal axes) are analysed to obtain the Fourier coefficients. The special terms have been calculated following a method indicated by R u n g e. The results are discussed in chapter VII.

When these Fourier coefficients are plotted in a Kohler diagram, characteristic functions are again found which are independent of the crystal, the impurity and the temperature.

CHAPTER I

GALLIUM, PROPERTIES AND TREATMENT;
ADJUSTING AND MEASURING-METHOD

Summary

After a general survey of the properties and treatment of gallium, the preparation of single-crystals, their adjusting and their examination are described in detail. The influence of the orientation of the crystallographic axes on the temperature dependence of the resistance has been investigated, and the results are discussed. The specific resistance at 0°C has been determined. The measuring method and the scheme of the measurements are explained.

1. *The crystallographic system.* The crystallographic system of gallium has been investigated by L a v e s [1]), who classified it in the rhombic system by means of Debije-Scherrer diagrams. Since the crystal axes in this system are orthogonal, it is possible to measure the properties of gallium in the direction of the different axes separately. This is an advantage over bismuth, where separate measurements are only possible in the direction of the principal axis. This is because the three binary axes of bismuth make angles of 120° with each other and therefore components of the effect in the direction of the two other axes affect measurements in the direction of the third.

It is important to notice that for gallium the shorter axes have the same length (according to L a v e s 4.51 ÅE, the longer axis being 7.51 ÅE; thus the axes ratio is 1.67 : 1 : 1). The different behaviour in the directions of the two shorter axes is due to a varying mutual arrangement of the atoms. Perpendicular to one of these axes the atoms lie flat-faced in a regular hexagon, whereas perpendicular to the other one the atoms lie alternately before and behind the average crystallographic plane, the hexagon in addition being somewhat deformed.

When we only consider the position of the lines in the Debije-

Scherrer diagrams, we are led to a tetragonal symmetry, as were initially J a e g e r, T e r p s t r a and W e s t e n b r i n k [2]). Therefore we usually refer to the longer axis as the pseudo-tetragonal one which for brevity we called the P_1-axis. L a v e s did not recognize the inequality of the shorter axes until he accounted for the intensity of the Debije-Scherrer lines.

The pseudo-tetragonal character is moreover demonstrated in the outward shape of the gallium crystals. Here only the ratio of the axes is important and so a full tetragonal symmetry is seen.

Employing a method of J a e g e r, T e r p s t r a and W e s- t e n b r i n k [2]) we produced single-crystals of gallium by putting a thin bar of glass, which was pre-cooled in liquid air, into some liquid metal, and then lifting it out again after a short time. Frequently we got in this way a truncated bi-pyramid, faced by the eight {111}-planes and two {001}-planes. The diagonal connecting the tops of the pyramides is the P_1-axis, the shorter axes are parallel to both the diagonals of a section transverse to the P_1-axis. For brevity we called the shorter axes the P_2- and P_3-axes. The specification will be explained more fully when discussing the results of the measurements.

2. *The temperature dependence of the electrical resistance.* Because of the pseudo-tetragonal structure of gallium, it may be expected that the behaviour of most of the physical properties will be scarcely distinguishable when measured in the direction of the two shorter axes, whereas there will be a remarkable difference with the behaviour in the direction of the longer axis.

We found that this was the case when measuring the temperature dependence of the electrical resistance in the direction of the three axes [3]). For that purpose we used three crystals (P_1-, P_2- and P_3-crystal) which had their lengths, coinciding with the direction of the current, parallel to the P_1-, P_2- and P_3-axes respectively. The preparation and the examination of the single-crystals, and the measuring method will be discussed elsewhere in this chapter. The temperature was measured by means of a platinum thermometer, which was calibrated beforehand by comparison with a helium gas thermometer.

It was assumed that the resistance is connected to the direction of the current by the formula:

$$R_{\alpha\beta\gamma} = R_1 \cos^2 \alpha + R_2 \cos^2 \beta + R_3 \cos^2 \gamma,$$

where $R_{\alpha\beta\gamma}$ is the resistance found when the current makes the angles α, β and γ with the P_1-, P_2- and P_3-axes respectively; R_1, R_2 and R_3 are the values of the resistance in the directions of the three axes. This is the simplest relation which may be expected following some considerations, introduced by V o i g t [4]), based on the symmetry of single-crystals. We did not investigate whether this formula is valid for gallium or not in this research, but for other metals it has been verified by previous experimentors.

According to Matthiessen's [5]) rule the electrical resistance of metals may be considered as the sum of a temperature independent part due to the impurities of the specimen and a temperature dependent part, generally known as the "ideal resistance", which would be found when measuring the resistance of an absolutely pure crystal.

Theoretically this is explained by assuming two sorts of irregularities in the crystal lattice, which do not influence each other:

1) perturbations due to the impurities (physical as well as chemical) of the crystal, causing the temperature independent part, and

2) perturbations due to the temperature movement of the atoms, causing the ideal part of the resistance.

At liquid helium temperatures the temperature dependent resistance is in general negligible compared with the temperature independent one, since it is increasing with a high (4th to 5th) power of the temperature in this range. The curve giving the resistance as a function of the temperature is therefore almost parallel to the temperature axis at liquid helium temperatures. By extrapolating this curve to the absolute zero point the temperature independent part of the resistance (for that reason called the residual resistance) may be determined and it is a measure of the impurity of the crystal concerned. By subtracting the residual resistance from the experimental value found at a given temperature the ideal resistance, R_{id}, connected with this temperature is found.

The experimental and the ideal resistance are further dependent on the dimensions of the crystal which are of course different for various specimens. For the comparison of the results it is therefore necessary to divide these experimental and ideal values by the corresponding values obtained at 0°C.

We have, therefore, values of $(R/R_{0°C})_{\text{id}}$ which are independent of the impurity and the dimensions of the crystals, and so the temperature curve of $(R/R_{0°C})_{id}$ gives for each metal a "characteristic" curve which shows the value 1 at 0°C.

Here again we have not investigated the validity of the previous assumptions for gallium.

With regard to the experimental results the following remarks can be made.

a. The P_1-crystal has at all temperatures a higher value of $(R/R_{0°C})_{id}$ than the others and is distinctly different from these two.

b. For the P_2- and P_3-crystals the values of $(R/R_{0°C})_{id}$ are much less different, and so it is found that the behaviour in the direction of the shorter axes is only just distinguishable. This made it impossible to draw in one figure two curves which could be seen separately. We were, however, more successful after dividing the values of $(R/R_{0°C})_{id}$ by the temperature. This may be seen in fig. 1. There is of course an intersection of the curves at 273°K, where the normalized value is 366.13×10^{-5}.

As G r ü n e i s e n [6]) showed, the temperature dependence of $(R/R_{0°C}T)_{id}$ may in general be assumed to have approximately the form of a Debije function. Although this is not strictly valid here, since the curves for the P_2- and P_3-crystals have a second intersection at 175°K (see c.), which can not occur for Debije curves, it nevertheless seems possible to consider the gallium curves over a certain range as Debije curves. The Θ-values (210°K for the P_1-, 230°K for the P_2-and 220°K for the P_3-crystal) are in good agreement with the one which can be estimated from some specific heat measurements (220°K).

c. At 175°K gallium behaves as if it were fully tetragonal in this respect, since the values of $(R/R_{0°C})_{id}$ are equal for the P_2-and P_3-crystals. Above 175°K the P_2-crystal shows the higher values, under 175°K this crystal shows in general the smaller values of $(R/R_{0°C})_{id}$.

d. The only exceptions on the rule noted under c. are found at the two lowest liquid hydrogen temperatures, where the P_2-crystal shows the higher values of $(R/R_{0°C})_{id}$ again. But, in the absence of liquid helium measurements for the P_2- and P_3-crystals, the determination of the residual resistances of these crystals has been relatively uncertain. Above 50°K this uncertainty has only a slight influence on the calculated values of $(R/R_{0°C})_{id}$; at liquid hydrogen temperatures it might, however, produce an important effect on the results.

It is for this reason that we did not draw special attention to these exceptions in an earlier publication on this subject [3]).

When discussing the results of our magnetic measurements we will, however, reconsider in more detail the possibility of a third intersection of the P_2- and P_3-curves near 25°K.

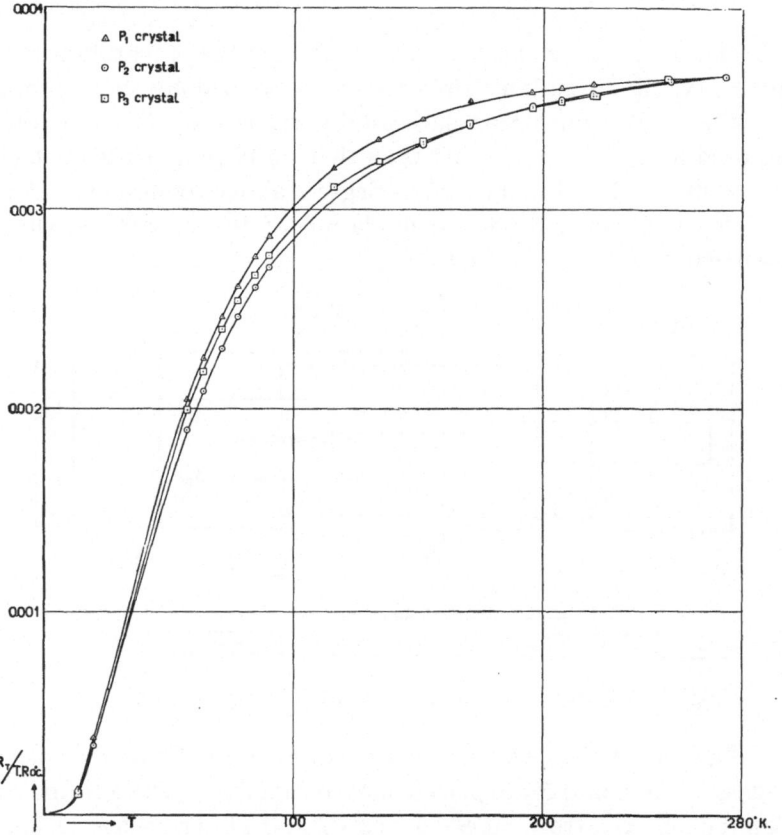

Fig. 1. Temperature dependence of the "ideal" resistance of single-crystals of gallium.

From the dimensions we have also calculated for the three crystals the specific resistance at 0°C. The accuracy of this calculation is not very great, since:

1) the length of the crystals was not greater than 1 cm because of the small extent of the homogeneous part of the magnetic field, and
2) the cross sections of the crystals were limited to about 1 mm²,

in order to get values of the resistances which were high enough for accurate measurements.

It is probably for these reasons that the values which were found were equal within the experimental error, giving:

$$\varrho_{0°C} = 5.27 \times 10^{-5} \, \Omega \, \text{cm.}$$

3. *The preparation of the single-crystals.* For the preparation of the single-crystals of gallium the method developed by S c h u b n i- k o w [8] for bismuth was successfully employed. The apparatus required for gallium is simpler than that for bismuth since the melt- ing point (30.2°C) lies only a few degrees above room temperature. The construction and the manipulation of the apparatus will be discussed with the help of fig. 2.

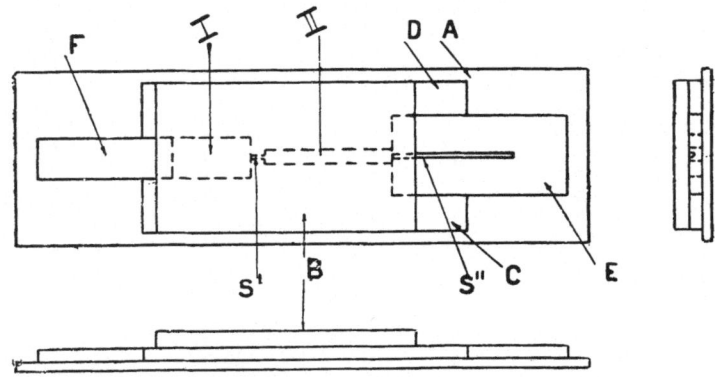

Fig. 2. Apparatus for the preparation of single-crystals.

A is a sheet of glass, which is placed horizontally. Four other sheets of glass, *C, D, E* and *F* are placed on it so that two cavities (*I* and *II*) are left open. A narrow slit *S'* connects *I* and *II*. The space *II* must have the form which is desired for the crystal. The sheets *C* and *D* are kept in the correct positions by means of two spring clips (not visible in the figure) which press them against the sheets *E* and *F*. The whole is then covered by the sheet of glass, *B*. All the glass sheets must be as flat as possible and fit together very well.

Before placing the cover sheet *B* in position some solid gallium is put into space *I*. Then after covering all with *B* the metal is melted by playing a spirit flame lightly under the base sheet *A*.

Now a seed crystal (usually one of the previously mentioned bipyramids) is fixed in some soft wax onto one of the faces of a copper cube. This cube is placed on sheet E in a position which depends on the orientation required in the crystal. Finally some liquid gallium is pushed through the canal S' in space II by means of sheet F. The liquid must be pushed so that a sufficient quantity of gallium comes out of the small groove S'' in sheet E to make contact with the seed crystal and melt it partly. When this has been attained the liquid will solidify during the gradual fall of temperature which then occurs, beginning with the seed crystal and thus continuing with its orientation.

Some disappointments which may occur are:

1) when the temperature of the liquid becomes too high the seed crystal will melt totally, and the crystallization which will begin after the cooling occurs at one (or more) arbitrary points with arbitrary orientation(s);

2) when the temperature of the liquid is not made high enough the entire seed crystal will remain solid, and it can be easily separated afterwards from the rest quite untouched. In this case too a crystal with one (or more) arbitrary orientation(s) will grow;

3) when the crystallization is proceeding correctly from the seed crystal a new centre may suddenly appear with an arbitrary orientation, so that a twin crystal is grown instead of a single one. With bismuth this happens frequently, the new centre occurring where some oxide is formed at the surface. To eliminate this some paraffin may be used which, after melting, spreads in a thin film over the liquid metal preventing oxidation and therefore twin growing. With gallium such a precaution is superfluous, since there is practically no oxidation at the (very low) melting point, and spontaneous crystallization only appears in exceptional cases because there is a strong sub-cooling effect.

In order to obtain the correct orientation of the seed crystal the following rules, which are inherently dependent on the type of crystal, can be made.

a. For the P_1-crystal the longer axis of the bi-pyramid (*i.e.* the P_1-axis) must be parallel to the length of space II, so that it will be parallel to the length of the crystal. For the other axes there are no essential conditions. But we preferred to have them perpendicular to the surfaces of the crystal so that their exact position was easily known. Therefore we carefully arranged to have one of the diagonals of the transverse section horizontal (the other one was then automatically vertical).

b. For the P_2- and P_3-crystals one of the diagonals of the transverse section of the bi-pyramid must be parallel to the length of space *II*. In addition we always preferred a horizontal or vertical position for the longer axis of the bi-pyramid in order to have it perpendicular to one of the surfaces of the crystal. Hence the remaining short axis is automatically perpendicular to the other side face.

For the preparation of the P_2- and P_3-crystals we often used as the seed crystal a discarded measuring crystal because of the simpler orientation.

4. *The examination of the single-crystals. a. External test.* When a gallium crystal is etched with hydrochloric acid the surface is covered with a great number of pits. The faces of these pits are crystallographic planes, thus forming "negative" crystals. When such an etched crystal is placed in a parallel pencil of light rays reflection only occurs in several well-defined directions. Such a reflection running continuously along the whole length of the specimen it proves the absence of twin orientations.

For a P_1-crystal in addition a periodicity of 90° must be found when it is rotated round its length.

Although it was not very probable that in the interior of the crystals there were any areas with different orientations, because of the small dimensions of our crystals, nevertheless an internal test was applied to most of them (the fact that a twin orientation in general grows from the surface suggests that some hidden orientation in the interior is not very likely).

b. Internal test. Following the method of D e H a a s and V a n A l p h e n [9]) Laue photographs were made by passing X-rays (with a continuous spectrum) through various parts of the crystal. For a single-crystal the Laue dots must lie on a straight line parallel to the lateral translation of the system crystal + film. Gallium has the advantage that there is less absorption of the X-rays on account of the low atomic number. The small transverse section of these crystals also diminishes the absorption so that an exposure of about 1 minute has in general been sufficient.

5. *The examination of the exact position of the axes.* This can be

done most easily from the results of the measurements of the resistance in a magnetic field. Since the planes through the crystal axes are symmetry planes, the values of ΔR must be equal for two directions of the field which are symmetrical with respect to them. Hence it follows that the rotational diagrams must be symmetrical both round the minimum and round the maximum when one of the axes is truly parallel to the length of the crystal and simultaneously the field is turned in a plane which is perpendicular to this length. In general a deviation of the axis which is supposed to be parallel to the length gives rise to an asymmetry in the diagrams. Only when the deviation happens to lie in a plane through the length and one of the axes perpendicular to it, does it escape detection.

The rotational diagrams can also help us to see whether the other axes are indeed perpendicular to the side faces of the crystal. For in that case the field must be perpendicular to these faces when a minimum or a maximum is found.

6. *The method of measuring the electrical resistance.* The electrical resistance has been measured by means of a Dieselhorst compensation apparatus. The unknown resistance is connected in a circuit with a standard resistance (normally 0,01 Ω) in series with it. The current density in this circuit was in general 100 mA. The potential differences between the ends of the unknown and of the standard resistances are compensated by using a second compensatory circuit. The required potential difference comes from the ends of a regulative resistance. The apparatus has the important advantage that the current in the compensation circuit remains constant while regulating this resistance. The ratio of the resistances necessary for the compensation of the corresponding potential differences equals the ratio of the unknown and the standard resistances. Thus the unknown may be calculated.

From what precedes it is seen that four leads are required, *i.e.* two current leads and two potential leads.

7. *The current and potential leads.* For the fusing of these leads to the crystal S c h u b n i k o w's method is again useful. The crystal and the lead are connected to the two ends of the secondary coil of a transformer so that they have a potential difference. The connection with the lead is made by means of a metal clip by which it is held

tight. On touching the crystal at the place where the joint is required, with the lead, a little spark is produced which causes a local
heating of the gallium. The voltage produced by the transformer
must be regulated in such a way that only a small quantity of the
gallium is melted. The copper lead is then pushed into the molten
metal and after that the clip must be removed as quickly as possible,
thus switching off the current. Since the surrounding metal takes
heat from the joint, the gallium solidifies again. Now, since this
metal, unlike most, expands on solidifying, the lead is held secure
thus ensuring a good electrical contact. The choice of the moment
when the lead must be pushed into the molten gallium requires some
experience. When it is done too early the lead can be removed easily
from the crystal; when it is done too late the end of the crystal is
melted away. The latter happens more frequently with gallium, because of its lower melting point, than with bismuth.

8. *The magnetic field.* A Weiss electromagnet was used with conic
pole pieces, which are flat-faced. The diameter of the ends was 40
mm; the gap was made 21 mm. In this gap a field, which is symmetrical round the common axis of the poles, appears when switching
on the electric current. The homogeneity of the field may be judged
from the fact that up to about 1 cm from the axis it is constant
within the experimental error ($1^o/_{oo}$).

The length of the crystals has for this reason been limited to 1 cm,
in order to be certain that the entire crystal is in the uniform part of
the field when only the middle of the crystal coincides with the
middle of the gap between the poles.

The magnetic field strength has been measured by the ballistic
method.

The whole magnet can be rotated about a vertical axis in such a
way that the homogeneous part of the field remains horizontal. The
position can be measured on a divided circle with an accuracy of $0.1°$.

9. *The temperature.* The outside diameter of the appendix of the
Dewar vessel containing the liquefied gases is 19 mm. So, when the
Dewar vessel is mounted between the poles of the electromagnet
there is a free space of 1 mm on each side of the glass.

The temperature was determined by means of the vapour pressure
(measured with a mercury manometer) of the liquids which were used,
i.e. ethylene, nitrogen, hydrogen and helium.

The constancy of the temperature could be controlled, especially at the higher temperatures, by measuring the resistance of the crystal itself, without a field. This control was made regularly every ten measuring points.

10. *The adjusting of the crystal. a.* When the effect is measured with the field always perpendicular to the current.

Since the field is always horizontal, the crystal must have a vertical position in this case. So it is bound on a vertical support of ivory, which itself is excentrically attached to a cylinder of copper. This is in turn fixed with DeKhotinsky cement on a long vertical rod of quartz, which is fastened in the top of the cryostat. The position of this rod must be carefully centred and the various distances must be arranged so that the crystal remains in the middle of the homogeneous part of the field when the magnet is rotated.

A deflection from the vertical causes an asymmetry in the rotational diagrams unless it happens to lie in one of the planes through the length and one of the axes perpendicular to the crystal. In this way, in general, deflections may be detected and then the position can be corrected.

b. When the influence on the effect is measured of β, the angle between the field and the current.

In this case the crystal is given a horizontal position, so that the angle β can be varied by turning the magnet. The support of ivory is now horizontal and in addition it may turn around a horizontal axis to enable us to correct the position of the crystal. This axis is perpendicular to the support and is at the lower end of a tube of German silver which is substituted in the place of the quartz rod. Through this tube a thin copper wire may be moved up and down, thus causing a rotation of the crystal in a vertical plane since the lower end of the wire is fastened to one of the ends of the ivory support. The upper end of the wire goes through an air-tight packing in the top of the cryostat and can therefore be handled from the outside.

11. *The purity of the material.* In view of the well-known fact that the magneto-resistance effect increases with the purity of metals S c h u b n i k o w developed a method for purifying bismuth [10]). Unfortunately gallium is too expensive for such a process, which has

a very small yield. Therefore the best we could realize was to select the purest gallium on the market. This proved to be the Hilger gallium (HS Brand), of which by chance two sorts were available in the laboratory. Messrs. Hilger's spectroscopic analysis declared one specimen much purer than the other, which had a greater amount of indium and zinc. The "pure" gallium showed, however, a much greater residual resistance than the "impure".

This may possibly be due to the fact that Zn lies with Ga in the same series of the periodic system but in the following column, while In lies in the same column with Ga, but in the following series. Now N o r b u r y's rule [11]) requires that the increase in resistance caused by a certain impurity (given in atomic %) increases with both the distance in the series of the periodic system and the distance in the columns in which the metal and the admixture lie. Thus, Zn and In being next neighbours of Ga in the periodic system, it may be expected that these metals contribute only a relatively small increase to the resistance, whereas much smaller admixtures of other metals exert a much greater influence.

In correspondence with experience the "pure" specimen of gallium showed together with the greater residual resistance, a smaller increase in resistance in a magnetic field at low temperatures, compared with the "impure". The conclusion seems to be that for crystals made from the "impure" material the perturbations in the lattice are less than for those made from the "pure". Therefore all the crystals used for the measurement of the magneto-resistance effect have been prepared with the "impure" gallium.

12. *Complications.* Two effects which may complicate the results of the measurements are the thermo-electric effect and the Hall effect. The influence of these effects is eliminated as follows.

1) *Thermo-electric effect.* The compensation apparatus is thermally insulated so effectively that no thermo-forces appear in it. By repeating every measurement after commutating simultaneously the currents in the measuring and compensation circuits the inevitable thermo-forces, due to the temperature differences at the joints on the crystal and on the leads to the compensation apparatus, may be eliminated, since on the average the effect vanishes.

2) *Hall effect.* The Hall effect, the potential gradient in a direction perpendicular to the plane through the current and the magnetic

field, is only perceptible when the line connecting the joints with the potential leads makes an angle with the length of the crystal. So it may be eliminated by carefully planting these joints in one of the edges of the crystal. That these joints have indeed been correctly made may be tested by repeating a measurement with the field reversed. If the same value of the resistance is found the Hall effect is absent. We have always succeeded in getting rid of this effect within the experimental error.

13. *The scheme of the measurements.* The measurements may be divided in two groups. In group I we determined the influence of the orientation of the field with respect to the crystal axes for the transverse effect (with the field perpendicular to the length and turning in a plane containing two axes). In group II the influence of the angle between field and current was considered. Both groups have been examined for P_1-, P_2- and P_3-crystals, thus six series of measurements were made. The results with the P_1-crystals are discussed in the chapters II (1st group) and IV (2nd group); those with the P_2- and P_3-crystals conjointly in the chapters III (1st group) and V (2nd group).

Each series is subdivided as follows:

1) determination of the rotational diagram (giving the influence of the orientation of the field) for a certain temperature and a certain field strength,

2) determination of the rotational diagrams for the same temperature but for various field strengths,

3) continuation of the measurements under 1) and 2) for various temperatures, and

4) measurement of the field dependences of the magneto-resistance effect for the particular positions where the field is parallel to any one of the axes.

REFERENCES

1) F. L a v e s, Naturwissenschaften **20**, 472, 1932; Z. Kristallograf. **84**, 256, 1933.
2) F. M. J a e g e r, P. T e r p s t r a and H. G. K. W e s t e n b r i n k, Proc. kon. Akad., Amsterdam **29**, 1193, 1926.
3) W. J. d e H a a s and J. W. B l o m, Commun. Kamerlingh Onnes Lab., Leiden No. 249c; Physica, 's-Grav. **4**, 767, 1937.
4) W. V o i g t, Lehrbuch der Kristallphysik (Teubner, Leipzig) 1910.
5) A. M a t t h i e s s e n and C. V o g t, Ann. Phys. Chem. (Pogg. Folge) **122**, 19, 1864.
6) E. G r ü n e i s e n, Verh. dtsch. phys. Ges. **15**, 186, 1913 and **20**, 36, 1918.
7) K. C l u s i u s and P. H a r t e c k, Z. phys. Chem. **134**, 243, 1928.
8) L. S c h u b n i k o w, Commun. No. 207b; Proc. kon. Akad., Amsterdam **33**, 327, 1930.
9) W. J. d e H a a s and P. M. v a n A l p h e n, Commun. No. 204d; Proc. kon. Akad., Amsterdam **33**, 128, 1930.
10) L. S c h u b n i k o w and W. J. d e H a a s, Commun. No. 207c; Proc. kon. Akad., Amsterdam **33**, 350, 1930.
11) A. L. N o r b u r y, Trans. Far. Soc. **16**, 570, 1921.

CHAPTER II

RESULTS OF THE MEASUREMENTS WITH THE P_1-CRYSTAL WHEN THE FIELD IS PERPENDICULAR TO THE CURRENT

Summary

The study of the increase in resistance in a magnetic field (ΔR) proves to be a sensitive method for the determination of slight differences in the atomic arrangement in the crystal lattice of gallium. The shorter axes, although having the same length, are clearly distinguishable with this method. At liquid nitrogen temperatures these axes are interchanged. It seems that the anomalous behaviour of gallium when the field is perpendicular to the so-called P_2-axis is responsible for this interchanging, as the transverse effect then shows an abnormal course for both the field and temperature dependencies. In the curves where ΔR is plotted as a function of the angle between the field and one of the shorter axes, secondary maxima and minima occur which have different characteristics for liquid nitrogen temperatures and liquid hydrogen temperatures. It is supposed that they have different origins too. At liquid helium temperatures the magnetic increase becomes constant and so does the normal resistance. This might be due to the fact that the magnetic increase depends on the irregularities in the lattice.

1. *Introduction.* As these measurements were published earlier [1] [2] [3]), only a general survey of the results obtained from the graphs is given here.

It was decided to start with the transverse effect for the P_1-crystal, since in this case the field is turned in the plane through both the short axes, and so it should be possible to trace a difference in behaviour between them.

The experience with other metals (even bismuth) induced us to begin with liquid hydrogen measurements, since in general in this temperature range the effect is greater than at higher temperatures.

2. *The rotational diagrams at 20.4°K.* The first measurements have been made at the boiling point of liquid hydrogen, 20.4°K [1]).

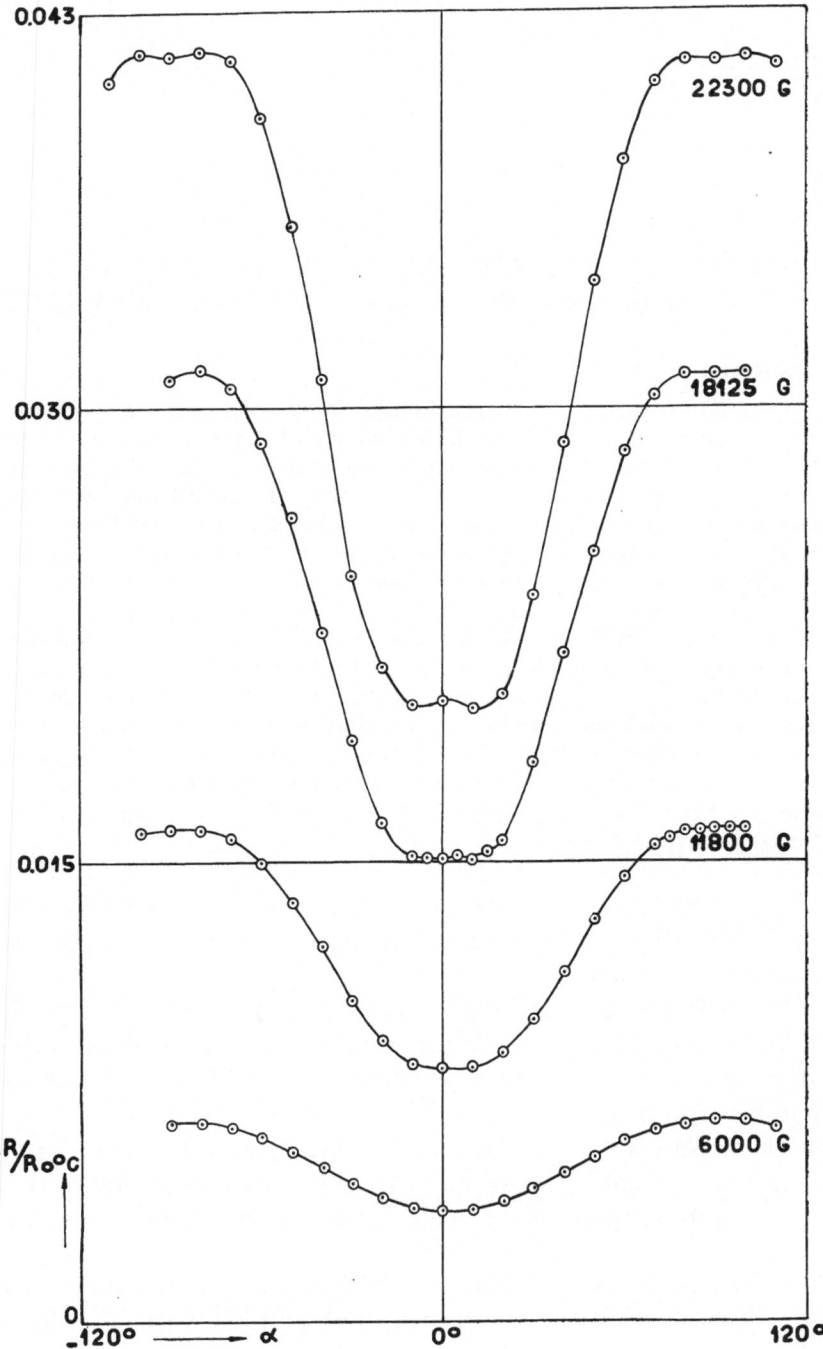

Fig. 1. Rotational diagrams for gallium at $T = 20.4°$K
P$_1$-crystal ($H \perp i$).

Fig. 1 shows the rotational diagrams giving $\Delta R/R_{0°C}$ as a function of a, the angle through which the field is turned from the position where a minimum increase of resistance is found at this temperature and in a weak field. The axis which, on account of symmetry considerations, is parallel to the field when a minimum is found ($a = 0°$), we defined as P_3-axis. The axis which is parallel to the field when a maximum is found in these circumstances ($a = \pm 90°$) was defined as P_2-axis.

The division by $R_{0°C}$ (the resistance in zero field at 0°C) was made to obtain results independent of the dimensions of the crystals. But K o h l e r's publications have afterwards shown that the same is achieved on ·dividing by R_{0T} (the resistance in zero field at the temperature of the measurement) and this would have been much more profitable for other reasons [4]).

It is obvious from the diagrams that there is an important difference between the values found when the field is parallel to the P_3-axis ($a = 0°$) and those for the field parallel to the P_2-axis ($a = \pm 90°$).

The ratio of both values is about 2, a pseudo-tetragonal symmetry therefore being out of the question altogether.

Thus, the magneto-resistance effect at low temperatures enables us to detect here small variations in the arrangement of the atoms, where the outward shape fails totally and the X-ray analysis only succeeds after an extensive research as with the temperature dependence of the electrical resistance.

Perhaps also for other metals, this effect, which seems to be highly sensitive to crystallographic differences, will help to determine the structure of the metal or the orientation of a single-crystal. In support of this suggestion may be mentioned:

1. the rotational diagrams found for bismuth [5]) show clearly the hexagonal structure of this metal;

2. measurements of J u s t i c.s. [6]) have shown that even for the cubic metals the orientation of the field has an important influence on the magnetic increase of the resistance at low temperatures;

3. in addition to the high sensitivity the very simple way of taking the measurements makes this method quite attractive.

Further results of the measurements at 20.4°K will be discussed together with those at the lower hydrogen temperatures.

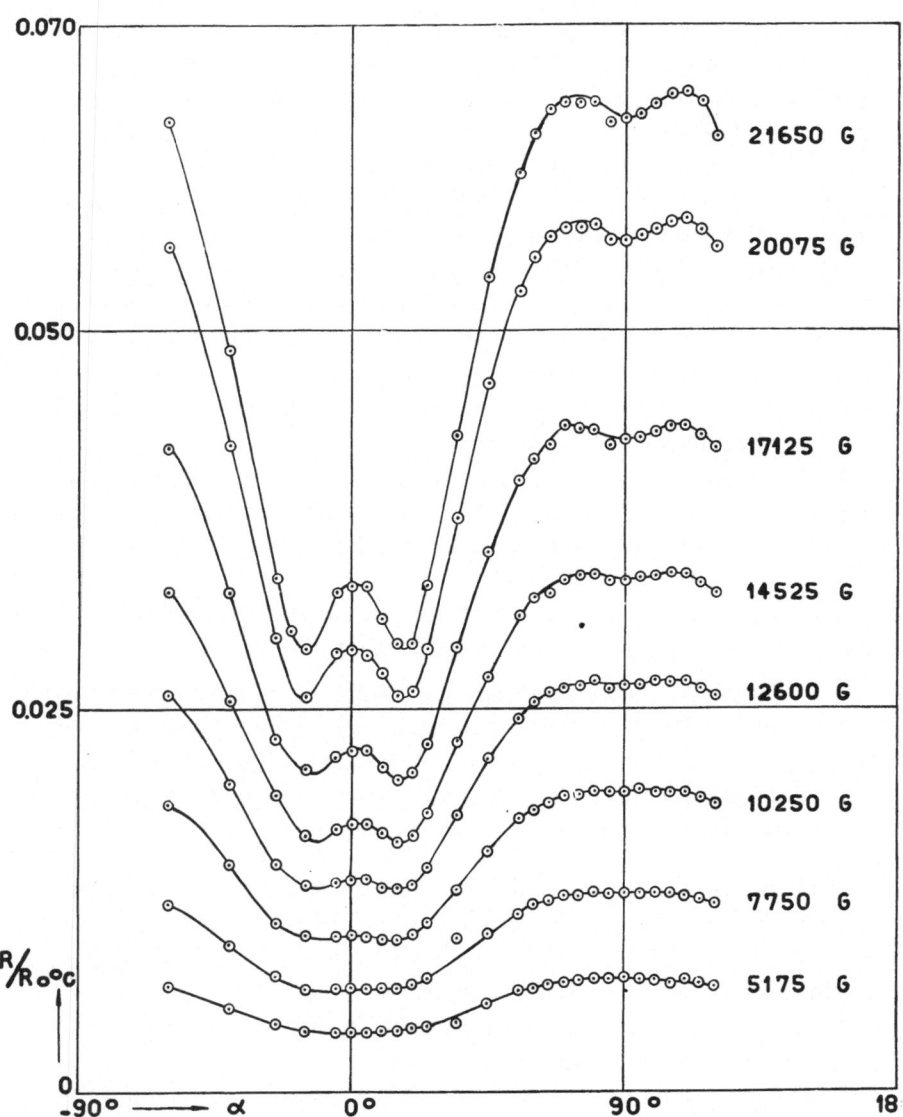

Fig. 2. Rotational diagrams for gallium at $T = 14.2°K$
P₁-crystal ($H \perp i$).

3. *The rotational diagrams at* 14.2°K, 10.4°K *and* 9.9°K. The experience with bismuth lead us to expect an intensification of some peculiarities, of which slight indications are already visible at the hydrogen boiling point, on proceeding to measurements at lower

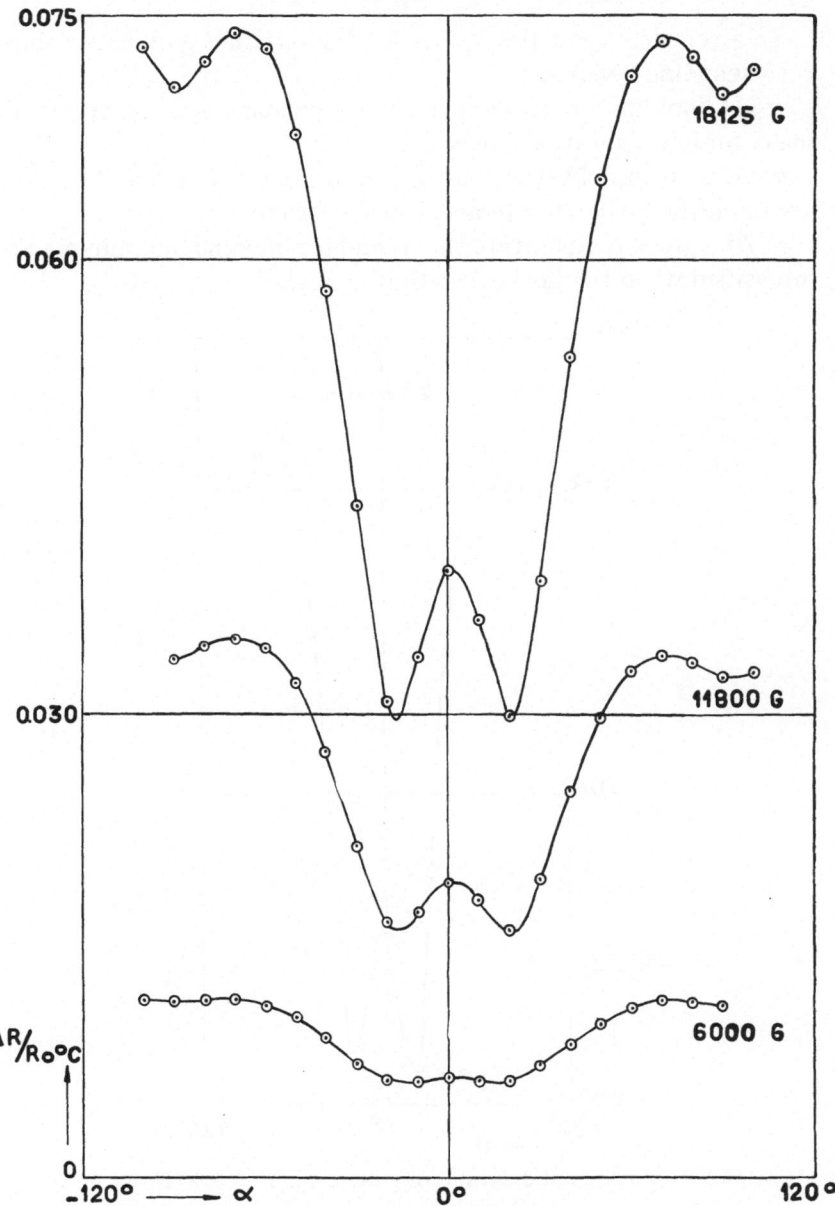

Fig. 3. Rotational diagrams for gallium at $T = 10.4°$K
P$_1$-crystal $(H \perp i)$.

temperatures. The diagrams are given in the figs. 2 $(T = 14.2°$K),
3 $(T = 10.4°$K) and 4 $(T = 9.9°$K).

To a certain extent the results for bismuth and gallium do show corresponding behaviour.

a. At liquid hydrogen temperatures secondary maxima and minima occur in the rotational diagrams.

b. At a given field strength these secondary maxima and minima are intensified when the temperature is lowered.

c. At a given temperature the secondary maxima and minima are intensified when the field strength is increased.

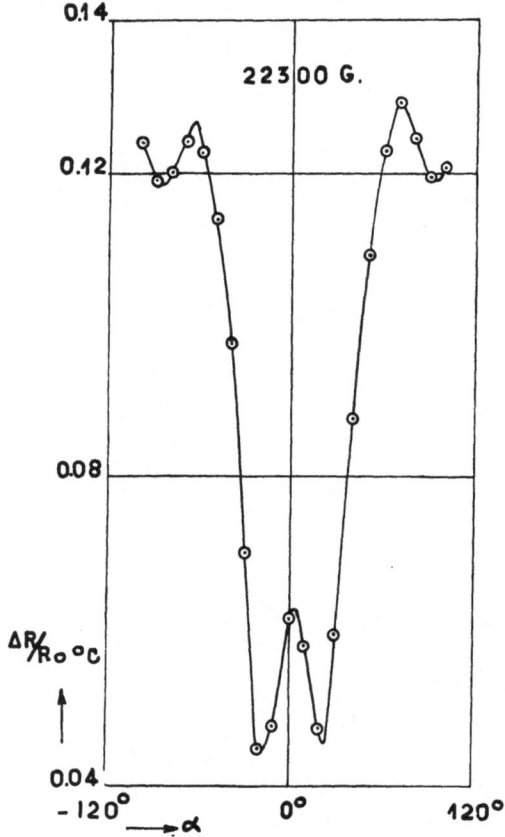

Fig. 4. Rotational diagram for gallium at $T = 9.9°$ K
P_1-crystal $(H \perp i)$.

That an increase of the field produces a change in the same direction as a lowering of the temperature is to be expected, since both have the same influence on the magnitude of $\Delta R/R_{0°C}$.

At the weaker fields, where the effect is smaller, only a flattening of the original maxima and minima is perceptible in general. From the rule mentioned under 2. it follows that at the lower temperatures this happens for a weaker field. The field strength at which the effect vanishes, so that a full sine-curve may be observed, also decreases with the temperature. At 20.4°K this is found for 6.00 kG, at 14.2°K it is barely reached at 5.175 kG, while at 10.4°K this field strength is below the experimental limit of the measurements (under 2 kG the measurements have lost all meaning, because of the inaccuracy which is inherent in the small values of $\Delta R/R_{0°C}$ corresponding to such a weak field).

On the other hand, however, some discrepancies are found between gallium and bismuth:

a. for bismuth a decrease in temperature is accompanied by the appearance of more and more complications in the diagrams, while for gallium only the existing effects increase, and

b. for bismuth the positions of the secondary maxima and minima are practically independent of temperature and field strength, while for gallium a certain dependency is found, which will be discussed in more detail when the liquid helium results are considered.

4. *The field dependence of $\Delta R/R_{0°C}$ at liquid hydrogen temperatures, with the field parallel to the P_2-axis and to the P_3-axis.* K a p i t z a [7]) has devised an ingenious method to overcome the difficulties of producing (for a short time) very strong fields, and has measured extensively the field dependence of the magnetic increase of the electrical resistance. He assumed the field dependency curve to consist, in general, of an initial parabolic part, which changes in a continuous manner to a linear variation with the field strength. The transition region connecting the parabolic with the linear part of the curve moves to the weaker fields either when the temperature is reduced or when a purer crystal is used.

It is noticeable that even in the linear part of the curve the value of ΔR remains unchanged when the field is reversed (for the parabolic part it is self-evident). When discussing the Hall effect we have already mentioned the fact that for gallium too this rule is valid.

For the P_1-crystals the field dependence of $\Delta R/R_{0°C}$ has been measured with the field parallel to the P_3-axis ($a = 0°$) and to the P_2-axis ($a = \pm 90°$) in this temperature range. The graphs in the

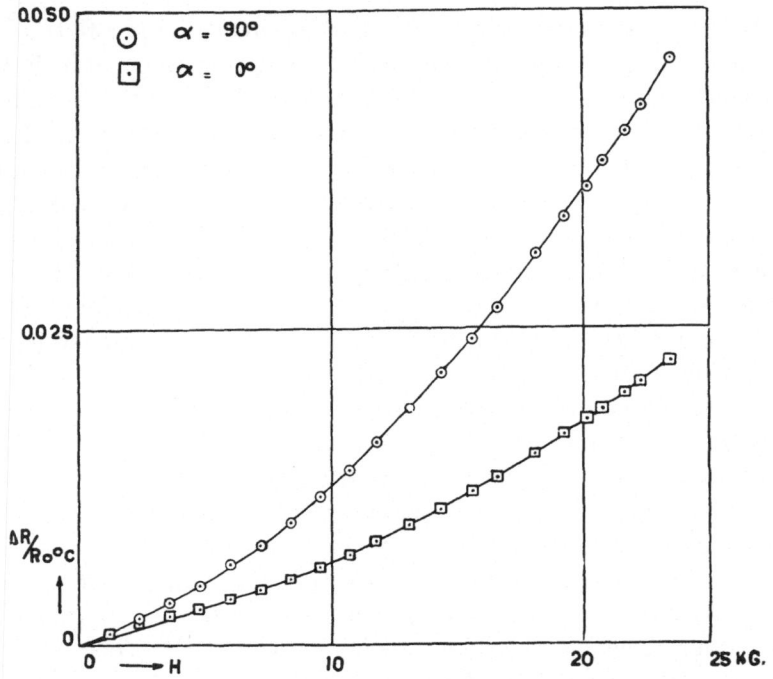

Fig. 5. Field dependence of $\Delta R_\perp/R_{0^\circ C}$ for gallium at $T = 20.4^\circ$K
P_1-crystal.

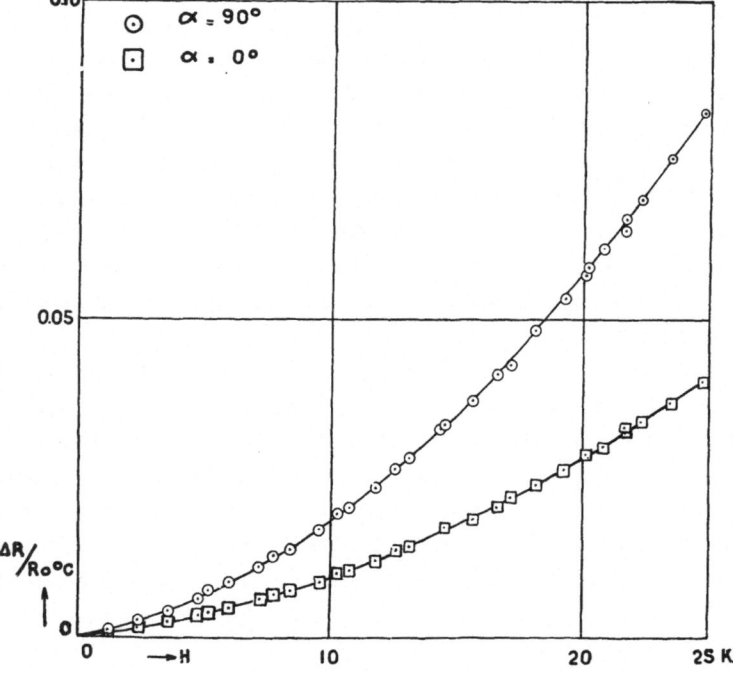

Fig. 6. Field dependence of $\Delta R_\perp/R_{0^\circ C}$ for gallium at $T = 14.2^\circ$K
P_1-crystal.

figs. 5 $(T = 20.4°K)$ and 6 $(T = 14.2°K)$ show the results. From the figures it will be seen that:

a. all the curves are fully parabolic, and there is no trace of the linear region found by K a p i t z a, and

b. the anomalies found in the field dependence curves of $\Delta R/R_{0°C}$ for bismuth in this temperature range do not appear with gallium.

5. *The measurements at liquid helium temperatures*. Although K a- p i t z a did not find the linear part in the field dependency curve of $\Delta R/R_{0°C}$ for all metals, and gallium might be one of these exceptional metals, it is nevertheless possible that the linear part may appear after:

1. using a purer crystal,

2. making the field stronger than the maximum available, and

3. lowering the temperature under the liquid hydrogen range. From the experience with bismuth it might also be expected that under these same conditions the anomalies both in the rotational diagrams, and in the field dependency curve of $\Delta R/R_{0°C}$, will perhaps appear for gallium as well.

Since a purification of the material was too expensive, and no stronger electromagnet was available for our experiments, our only chance was to continue the measurements at liquid helium temperatures and perhaps find by that means the missing effects.

Before the discussion of our gallium results we give a short report on our measurements [8]) with the purest of S c h u b n i k o w's bis- muth crystals at liquid helium temperatures. In fig. 7 the diagrams at liquid helium temperatures are given, in fig. 8 the results of the control measurements at 14.15°K. In fig. 9 are drawn the field depend- ency curves of $\Delta R/R_{0°C}$ with the field parallel to one of the binary axes $(\varphi = 0°)$ and with the field perpendicular to them $(\varphi = \pm 30°)$. For comparison the results found at 14.15°K are also given.

In summary the following may be concluded.

a. In accordance with the expectations we find at 4.22°K the ano- malies in the curves more pronounced than at 14.15°K. Where at 14.15°K we only see a slight indication of a singularity, at 4.22°K we find a secondary maximum or minimum, and consequently the posi- tions of these maxima or minima are almost independent of temper- ature.

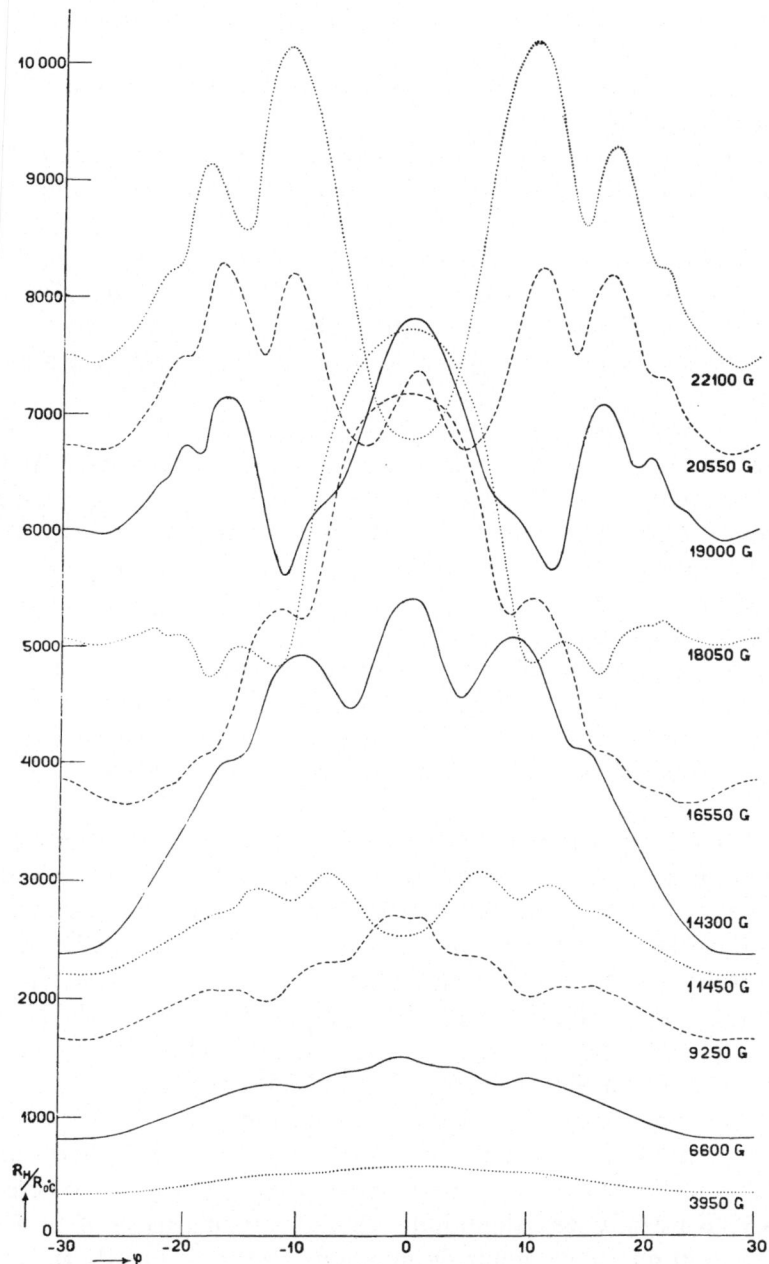

Fig. 7. Rotational diagrams for bismuth at $T = 4.22°$K.

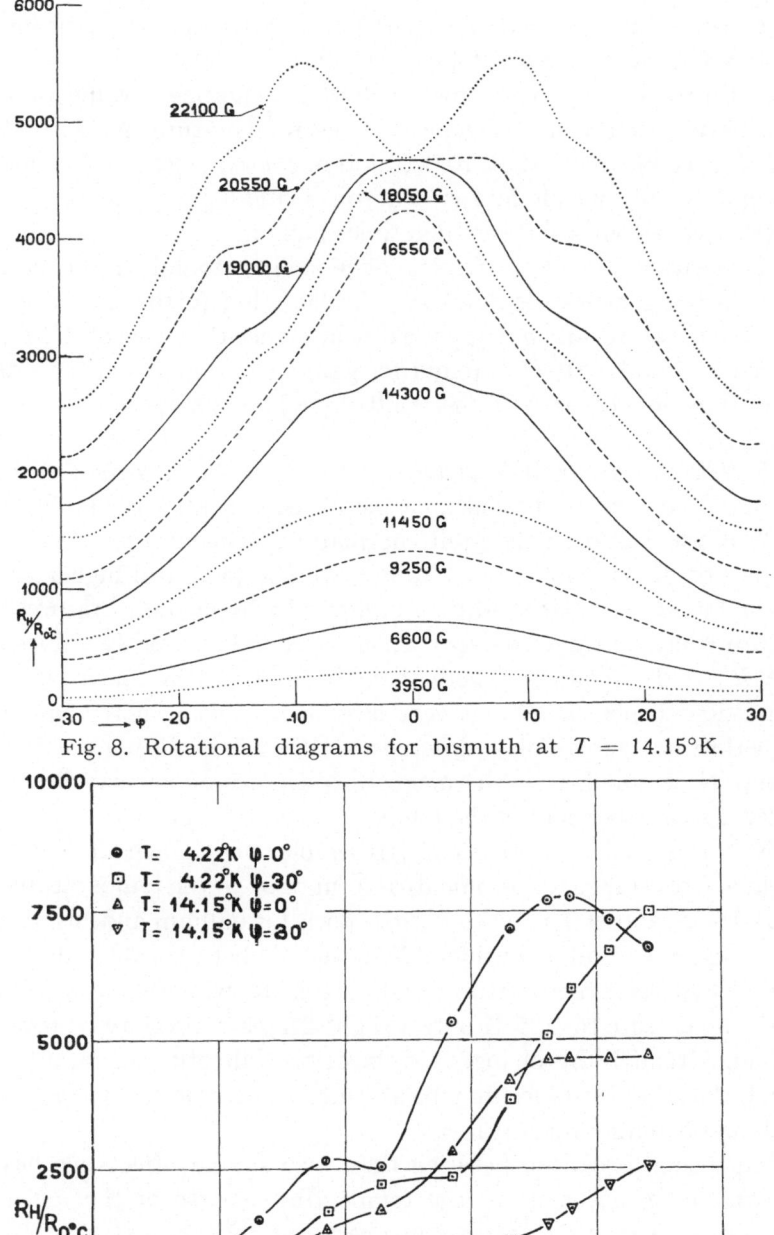

Fig. 8. Rotational diagrams for bismuth at $T = 14.15°$K.

Fig. 9. Field dependence of $\Delta R_\perp / R_{0°C}$ for bismuth.

b. Since the material was very pure, the resistance of bismuth in zero field is very small at 4.22°K ($R/R_{0°C} = 0.003$).

c. Corresponding with this we find an enormous value of the magnetic increase in resistance at this temperature. At 22.10 kG $\Delta R/R_{0°C}$ can be more than 10 000, and in consequence $\Delta R/R_{0T}$ more than 3×10^6, which means that on switching on the field the resistance becomes three million times higher.

d. Perhaps the most striking phenomenon found at the liquid helium temperatures is, however, the fact that further lowering of the temperature does not appreciably influence the value of $\Delta R/R_{0°C}$.

Since a similar effect is found for gallium too, it will now be discussed in more detail with the results found for this metal.

6. *The temperature independence of $\Delta R/R_{0°C}$ at liquid helium temperatures.* The decision to continue the measurements with gallium to 1.35°K was based on the (quite normal) experimental fact that the values of $\Delta R/R_{0°C}$ found at 4.22°K [2]) were a good deal higher than those found in similar conditions at liquid hydrogen temperatures.

However, contrary to expectation we found at 1.35°K values of $\Delta R/R_{0°C}$ only about 1% more than those at 4.22°K under similar conditions (since $\Delta R/R_{0°C}$ is very large here, this is more than the experimental error). These differences, although in the normal direction, are so small, that the magneto-resistance effect may be considered as independent of the temperature in this range.

Now this behaviour of $\Delta R/R_{0°C}$ resembles very much that of the ordinary resistance (in zero field) as a function of the temperature in this range, and this is quite normal both for gallium and bismuth. Here again we find a considerable difference between the values at liquid hydrogen temperatures and at 4.22°K, whereas on lowering the temperature to 1.35°K only a slight decrease of the resistance is found. Although this change is in the correct direction again, it is so small that the resistance may be assumed to be practically constant at liquid helium temperatures.

Supposing a relation between these two similar effects we have drawn in fig. 10, next to the temperature course of the normal resistance, that of the increase in a certain field (22.30 kG) both for $a = 0°$ and for $a = \pm 90°$. Since the normal resistance is increasing when the temperature is raised, whereas $\Delta R/R_{0°C}$ is decreasing, we decided to plot the reciprocal values of $\Delta R/R_{0°C}$ (*i.e.* $R_{0°C}/\Delta R$). This makes the comparison easier and shows better the similarity between both these temperature dependencies. In order to get the desired value of $\Delta R/R_{0°C}$ at liquid helium temperatures it was necessary to extrapolate the field dependency curves, for these were only known up to 22.10 kG, and we required the value at 22.30 kG. (The maximum field strength in this temperature range was limited to 22.10 kG because a second Dewar vessel containing liquid hydrogen had to be used and so the gap between the magnet poles was increased).

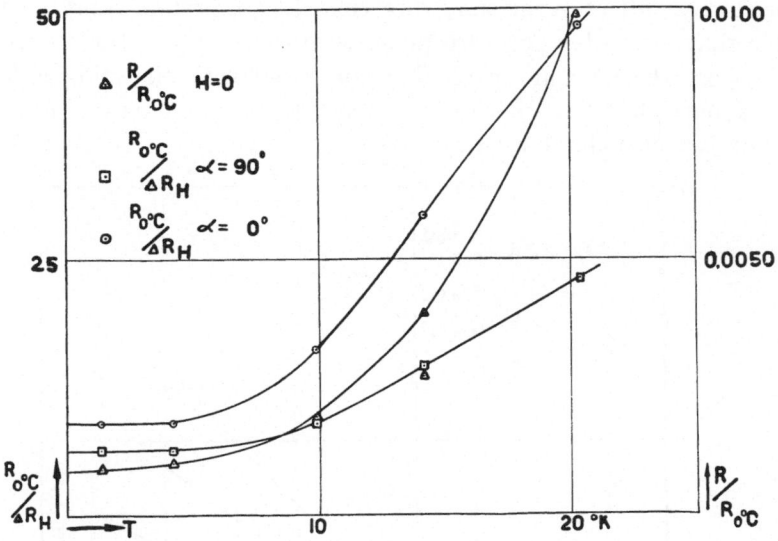

Fig. 10. Temperature dependence of the normal resistance and of the reciprocal of the magnetic increase of resistance at 22.30 kG.

It is evident from the graph that the resistance and the magnetic increase tend to become constant in the same temperature range. This assumption is supported by the results found by D e H a a s and V a n A l p h e n [9]) with the alloys Cu-Zn (70% Zn) and Cd-Hg (30% Hg). For both alloys the normal resistance is, as usual at low temperatures, independent of the temperature, and in addition under 77°K no change in the value of $\Delta R/R_{0°C}$ is found.

Combining this analogy of the behaviour of the resistance and of the magnetic increase with the well known fact that a purer material shows a smaller residual resistance and also a greater magnetic increase (the validity of this rule for gallium has been mentioned previously), one may, on the basis of the modern theories of the electrical conductivity, come to the following conclusion:

An increase of the perturbations in the crystal lattice is combined with a decrease of the influence of a magnetic field on the resistance.

In addition it does not matter whether it is the temperature dependent or the temperature independent irregularities which are growing. So purification of the material and lowering of the temperature both produce an effect in the same direction. It is therefore comprehensible that in the temperature range where the temperature dependent perturbations are negligible compared with the independent ones, both the resistance and the magnetic increase are nearly constant (see also chapter VI).

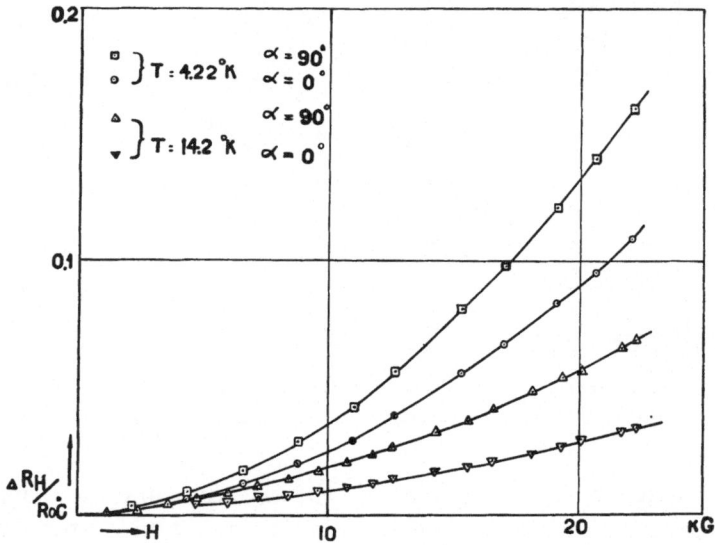

Fig. 11. Field dependence of $\Delta R_{\perp}/R_{0°C}$ for gallium.

7. *The field dependence of $\Delta R/R_{0°C}$ with the field parallel to the P_2- and to the P_3-axes and the rotational diagrams at liquid helium temperatures.* In fig. 11 the field dependences of $\Delta R/R_{0°C}$ are given at 4.22°K with the field parallel to the P_3-axis ($a = 0°$) and parallel to the P_2-axis ($a = \pm 90°$). The results found at 1.35°K are omitted, since the slight differences between the measurements at different liquid helium temperatures make it impossible to distinguish between the

curves in one figure. The results found at 14.15°K are given once
more to show that the liquid helium values of $\Delta R/R_{0°C}$ are many
times greater than those found at the liquid hydrogen temperatures.
Fig. 12 shows the rotational diagrams at 1.35°K, the results at
4.22°K being omitted for the same reason as above.

a. From the graphs it is evident that neither the linear part in the
field dependency curve of $\Delta R/R_{0°C}$ (investigated by K a p i t z a for
other metals), nor the anomalies in the rotational diagrams and in
the field dependency curve of $\Delta R/R_{0°C}$ found for bismuth (by
S c h u b n i k o w) appeared on lowering the temperature, contrary
to our expectations.

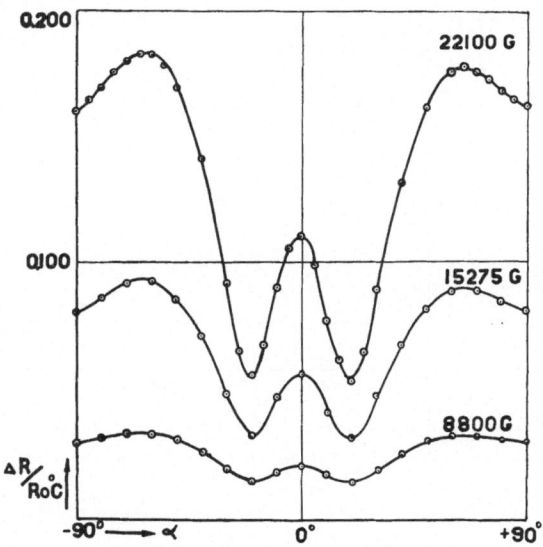

Fig. 12. Rotational diagrams for gallium at $T = 1.35°$K
P_1-crystal $(H \perp i)$.

The temperature independency of the measurements in this range
makes it certain that further lowering of the temperature cannot
give rise to the appearance of these effects. For this reason we regret
very much that the circumstances have prevented us from employing both the other means which could have helped.

b. The secondary maxima and minima in the rotational diagrams
are intensified at liquid helium temperatures compared with those
in the liquid hydrogen region as might be expected. We gave special
attention to the positions of the secondary maxima on either side of

the original minimum and of the minima on either side of the original maximum. The liquid hydrogen results led us to suppose that these positions might depend both on the temperature and on the field strength. For the comparison all the distances between these maxima and minima have been collected in tables I (separations of the secondary minima) and II (separations of the secondary maxima) for both the liquid hydrogen and helium ranges.

TABLE I

H kG			Separation of the secondary minima.		
	$T = 20.4°K$	$T = 14.2°K$	$T = 10.4°K$	$T = 9.9°K$	$T = 4.2°K$
6.000			31°		
8.800					38°
10.250		24°			
11.800			37		
12.600		26			
14.525		28			
15.275					40
17.125		29			
18.125			38		
20.075		32			
21.650		33			
22.100					40
22.300	19°			41°	

TABLE II

H kG			Separation of the secondary maxima.		
	$T = 20.4°K$	$T = 14.2°K$	$T = 10.4°K$	$T = 9.9°K$	$T = 4.2°K$
6.000			34°		
8.800					50°
10.250		24°			
11.800			36		
12.600		26			
14.525		28			
15.275					50
17.125		34			
18.125			38		
20.075		33			
21.650		35			
22.100					52
22.300	21°			45°	

These separations do indeed seem to be dependent on the temperature and on the field strength. Lowering of the temperature and increase of the field increase these distances and act, as usual, in the same direction.

8. *The rotational diagrams at* 49.8°K. The results found for gallium at liquid hydrogen temperatures, together with those for bismuth, led us to expect full sine-curves for the rotational diagrams at liquid nitrogen temperatures. This can be easily understood when one remembers that at liquid hydrogen temperatures the secondary maxima and minima vanish when the temperature is raised, so that at 20.4°K they only exist with 22.30 kG, while for the other fields only a slight flattening is perceptible. So, since it seemed highly probable that the anomalies would be absent, we intended initially to measure only a few diagrams in this temperature range, thus completing our results. The experience, however, was by no means in accordance with this line of thought and so an extensive research was necessary [3]).

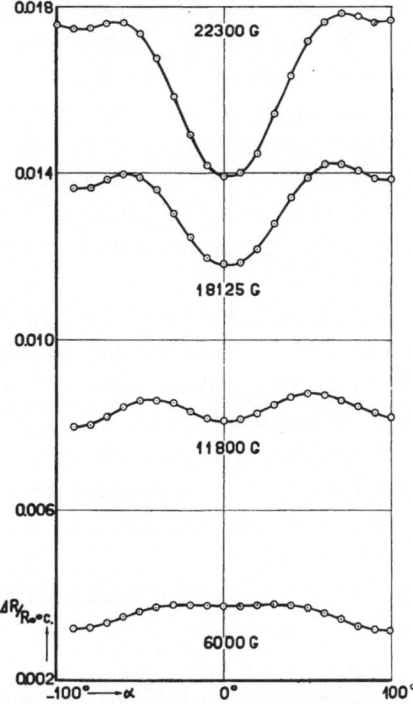

Fig. 13. Rotational diagrams for gallium at $T = 49.8$°K
P_1-crystal ($H \perp i$).

Since it is possible to see all the essential particularities of the rotational diagrams in this temperature region by considering the measurements at the lowest nitrogen temperature (solid nitrogen) we decided to begin the discussion with the results at 49.8°K. The diagrams are given in fig. 13.

Considering the graphs the following may be noticed.

a. From the diagram at 6.00 kG it is at once evident that we do not have a full sine-curve here, whilst at 20.4°K for this field we do.

b. It can be seen that the original minimum is undisturbed, whereas the original maximum has been replaced by a secondary minimum. Hence it follows that the secondary minimum in this temperature range and the corresponding anomalies at the liquid hydrogen temperatures must have different origins.

c. The most important singularity may only be noticed by comparing the diagram found at this temperature with that for the same field, 6.00 kG, in the liquid hydrogen range. Then one may see that at 49.8°K the minimum is found when the field is parallel to the P_2-axis ($a = \pm 90°$), while at 20.4°K for the same field it is found when the field is parallel to the P_3-axis ($a = 0°$). Thus on lowering the temperature the shorter axes interchange their roles.

d. A similar interchanging appears on increasing the field, as may be seen from the comparison of the diagrams at 18.125 kG and 22.30 kG with the diagram at 6.00 kG. For the stronger fields the smallest value of $\Delta R/R_{0°C}$ is found with the field parallel to the P_3-axis ($a = 0°$), whereas for 6.00 kG it is found when the field is parallel to the P_2-axis ($a = \pm 90°$).

e. The secondary minimum replacing the original maximum at $a = 0°$ is more pronounced at 22.30 kG than at 18.125 kG, while at 6.00 kG it is only just perceptible. Assuming the depth of the secondary minimum under influence of the growing field to increase continuously, it may be expected that somewhere between 6.00 kG and 18.125 kG a field exists where the secondary minimum equals the original minimum at $a = \pm 90°$. This indeed is found at 11.80 kG, where a periodicity of 90° may be seen for this reason in the rotational diagrams. In this condition gallium behaves in a fully tetragonal manner.

9. *The rotational diagrams at* 56.5°*K,* 70.2°*K and* 77.4°*K.* These diagrams are given in the figs. 14 ($T = 56.5°$K), 15 ($T = 70.2°$K) and 16 ($T = 77.4°$K). The results at these higher temperatures in the liquid nitrogen region confirm in general those found at 49.8°K. This may be illustrated by a choice of examples.

a. The rotational diagrams are never fully sinusoidal in this temperature range, contrary to expectations.

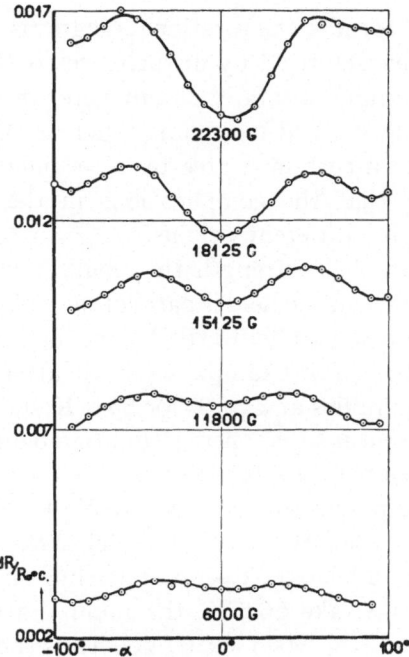

Fig. 14. Rotational diagrams for gallium at $T = 56.5°$K
P$_1$-crystal ($H \perp i$).

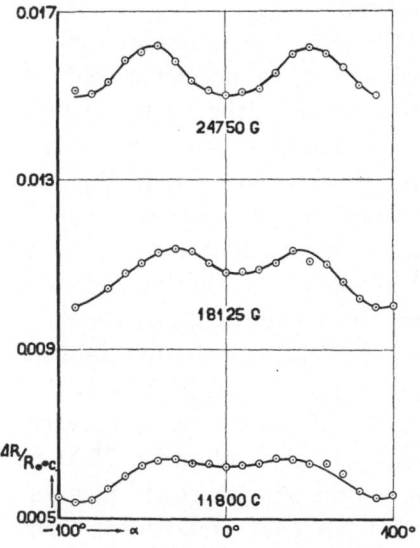

Fig. 15. Rotational diagrams
for gallium at $T = 70.2°$K
P$_1$-crystal ($H \perp i$).

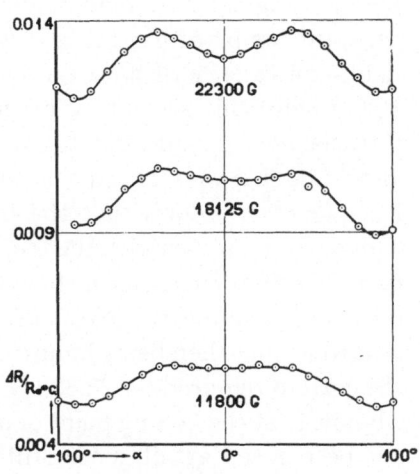

Fig. 16. Rotational diagrams
for gallium at $T = 77.4°$K
P$_1$-crystal ($H \perp i$).

b. The special form of the rotational diagrams at liquid hydrogen temperatures, characterized by disturbances both in the minimum and in the maximum, is totally absent here. In this range there is invariably an undisturbed minimum, whereas the maximum is always replaced to a certain degree by a secondary minimum. It is highly probable that the complications in the two temperature regions have quite different origins.

c. For a certain field strength the smallest value of $\Delta R/R_{0°C}$ is found sometimes when the field is parallel to the P_3-axis ($\alpha = 0°$) and sometimes when it is parallel to the P_2-axis ($\alpha = \pm 90°$) depending on the temperature. For example we draw attention to 22.30 kG, where the minimum lies at $\alpha = 0°$ for 56.5°K and at $\alpha = \pm 90°$ for 77.4°K. This rule means that there is an interchanging of the shorter axes when the temperature is varied in this range.

d. For a certain temperature the smallest value of $\Delta R/R_{0°C}$ is found sometimes when the field is parallel to the P_3-axis and sometimes when it is parallel to the P_2-axis depending on the field strength. For example, if we take 56.5°K, the minimum lies at $\alpha = 0°$ for 22.30 kG and at $\alpha = \pm 90°$ for 6.00 kG. Thus the interchanging of the shorter axes may also be influenced by a variation of the field. As usual the same effect can be obtained either by an increase of the field or by a decrease of the temperature.

e. Assuming that the deepening of the secondary minimum replacing the original maximum under influence of an increasing field (at constant temperature) takes place continuously, then, for a certain field, equivalence of both the minima is required, since one knows that finally this secondary minimum becomes lower than the undisturbed one. In addition to the previously mentioned case (49.8°K and 11.80 kG) we realized this condition also for 56.5°K and 15.125 kG. Hence it follows that the field at which the shorter axes are equivalent increases with the temperature. Therefore it was necessary at 70.2°K to increase the field up to 24.75 kG in order to establish tetragonal symmetry. (Normally we did not exceed 22.30 kG, the electromagnet then being almost saturated. However, by using twice the current required for 22.30 kG it was possible to obtain 24.75 kG). Obviously at the boiling point of nitrogen the periodicity of 90° could not be reached at all, since still a stronger field would have been necessary.

f. The converse of the rule mentioned in *e.* also occurs. This means that for a certain field a temperature exists at which the shorter axes are equivalent and this temperature increases with the field. For 6.00 kG we only know that this temperature lies between 20.4°K and 49.8°K, since the maximum and minimum interchange on passing from one of these temperatures to the other. Because of the continuity the shorter axes must have been equivalent at some intermediate temperature. For 11.80 kG, 15.125 kG and 24.75 kG the periodicity of 90° is found at 49.8°K, 56.5°K and 70.2°K respectively. For 18.125 kG and 22.30 kG this temperature must lie between 56.5°K and 70.2°K.

g. Finally it can be stated that the general rule requiring an increasing value of $\Delta R/R_{0°C}$ with falling temperature is perfectly obeyed in this range.

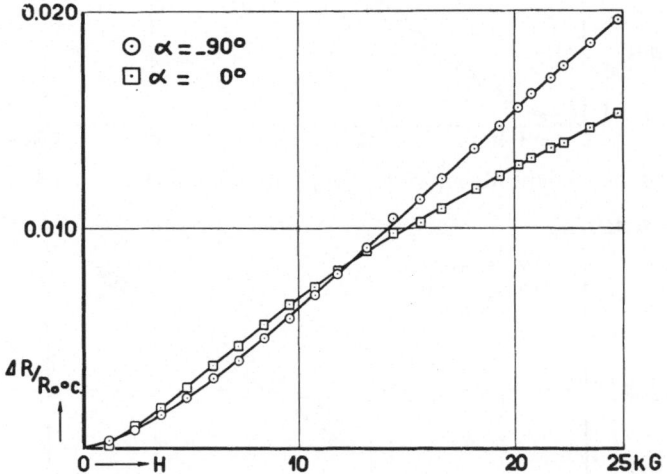

Fig. 17. Field dependence of $\Delta R_{\perp}/R_{0°C}$ for gallium at $T = 49.8°K$.

10. *The field dependence of $\Delta R/R_{0°C}$ with the field parallel to the P_2- and P_3-axes at liquid nitrogen temperatures.* The graphs are given in the figs. 17 ($T = 49.8°K$), 18 ($T = 56.5°K$), 19 ($T = 70.2°K$) and 20 ($T = 77.4°K$). With the field parallel to the P_2-axis ($a = \pm 90°$) a normal parabolic curve is found at all temperatures and as at liquid hydrogen (and even at liquid helium) temperatures there is again no linear part. With the field parallel to the P_3-axis ($a = 0°$), however, an anomalous behaviour is found for the field dependence of $\Delta R/R_{0°C}$. At all liquid nitrogen temperatures we find that the normal parabolic

beginning changes continuously into a linear region. This linear part is very short and is followed afterwards by a curve which has a gradually decreasing slope. The further course of the curve at stronger fields remains unpredictable.

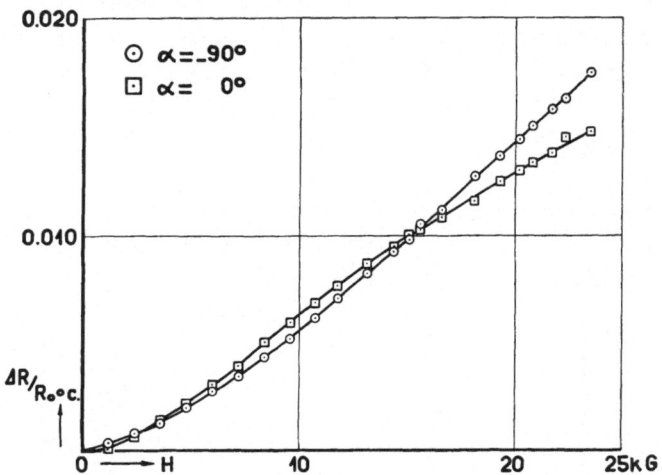

Fig. 18. Field dependence of $\Delta R_\perp/R_{0°C}$ for gallium at $T = 56.5°$K.

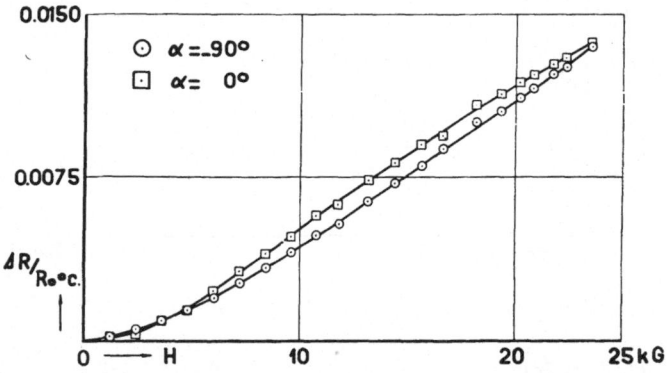

Fig. 19. Field dependence of $\Delta R_\perp/R_{0°C}$ for gallium at $T = 70.2°$K.

Because of this behaviour the anomalous curve, which at the weaker fields shows the higher values of $\Delta R/R_{0°C}$, tends towards the normal one. When the initial difference (which is relatively small in this temperature region on account of the pseudo-tetragonality) is small enough an intersection of both curves follows. The corresponding values of temperature and field strength for which this intersection occurs must evidently coincide with those found with the rotational diagrams, since in both cases the equivalence of the shorter axes is manifestated. We found that this was so (as may be seen from the values: 49.8°K and 12.0 kG, 56.5°K and 15.5 kG and 70.2°K and 24.0 kG). At the boiling point of nitrogen the intersection lies at a field above the limit which our electromagnet can produce.

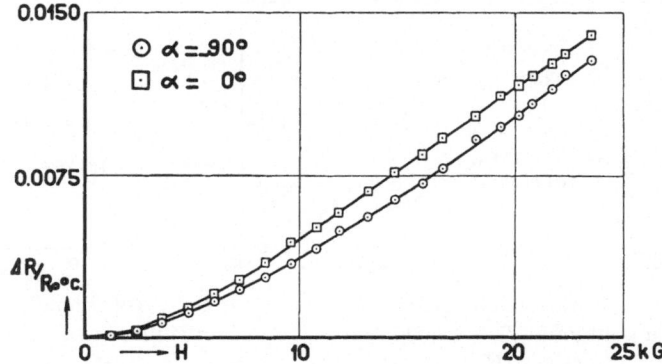

Fig. 20. Field dependence of $\Delta R_{\perp}/R_{0°C}$ for gallium at $T = 77.4°K$.

The corresponding values of temperature and field strength have been plotted in a graph, and the resulting curve has been extrapolated to zero field. Thus a rough estimation of the temperature at which gallium would behave in a tetragonal manner in zero field (with respect to electrical properties) was found. This approximation shows that the equivalence of the shorter axes (in zero field) lies at about 30°K.

With regard to the previously mentioned conclusion that the curves, giving the ideal resistances of the P_2- and P_3-crystals as a function of the temperature, intersect not only at 273°K and 175°K but also at about 25°K, it seems probable now that it is less questionable than we initially supposed. The agreement between these two temperatures (25 and 30°K) at which tetragonal behaviour occurs is quite surprising considering the inaccuracy of the determinations.

The result of our investigation of a similar coincidence on extrapolating to a zero field value near 175°K can only be given after the discussion of our measurements at higher temperatures.

11. *The rotational diagrams at liquid ethylene temperatures.* Now, since $\Delta R/R_{0°C}$ diminishes when the temperature is raised so that the

magneto-resistance effect is relatively small in this region, especially for the weaker fields, we confined the measurements to the diagrams for the stronger fields (18.125 kG and 22.30 kG) where the accuracy is sufficiently high. The graphs are given in the figs. 21 ($T = 126°$K) and 22 ($T = 155°$K).

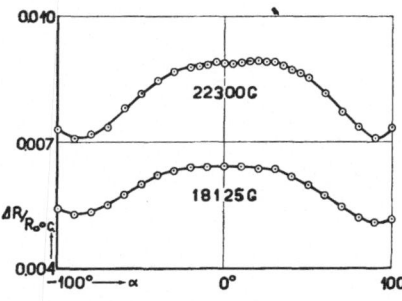

Fig. 21. Rotational diagrams
for gallium at $T = 126°$K
P_1-crystal ($H \perp i$).

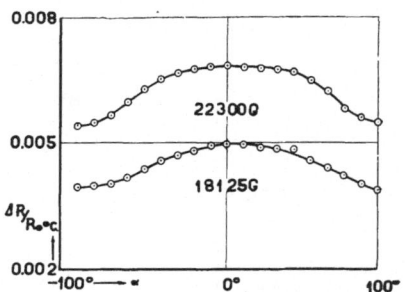

Fig. 22. Rotational diagrams
for gallium at $T = 155°$K
P_1-crystal ($H \perp i$).

From the results the following points may be noticed.

a. As can be expected from the results at liquid nitrogen temperatures, a minimum is found in the diagrams when the field is parallel to the P_2-axis ($a = \pm 90°$). This is just the opposite behaviour to that found at liquid hydrogen temperatures when it appeared with the field parallel to the P_3-axis ($a = 0°$). Now the shorter axes have completely interchanged.

b. The decreased depth of the secondary minimum replacing the original maximum at $a = 0°$ may also be expected from the results of the liquid nitrogen diagrams. At 126°K a flattening of the maximum indicates the disturbance; at 155°K this is only perceptible for 22.30 kG, whereas for 18.125 kG an almost full sine-curve is found.

c. It is obvious from these measurements that the equivalence of the P_2- and P_3-axes near 175°K (see section 10) cannot be concluded from an extrapolation to zero field.

d. Finally we have tried to continue the measurements to still higher temperatures.

a. We hoped at first to do this by means of liquid methylchloride. But the temperature fluctuations caused alternately by the strong sub-cooling and violent bumping seriously affect the measurement, by producing variations in the resistance, which may even be of the same order as those due to the magnetic field. The temperature variation during the time required for the measurement of a single point often caused an apparent decrease in resistance. Unfortunately the smallness of $\Delta R/R_{0^\circ C}$ even at the maximum field strength makes this effect of even more consequence. Because of these irregularities it was impossible to draw continuous curves through the plotted points.

β. Secondly we tried to measure the rotational diagrams at room temperature. Here, however, the variations of the resistance connected with a rotation of the electromagnet are so small that they lie within the experimental error.

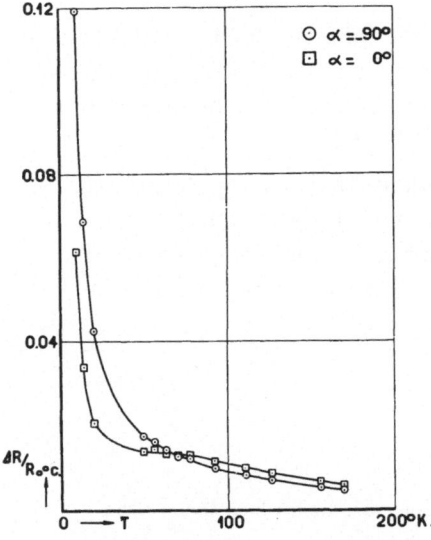 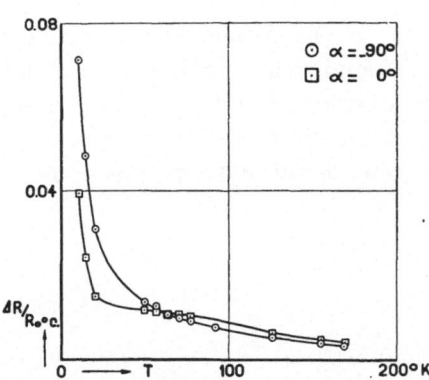

Fig. 23. Temperature dependence of $\Delta R_\perp/R_{0^\circ C}$ for gallium at $H = 22.300$ kG.

Fig. 24. Temperature dependence of $\Delta R_\perp/R_{0^\circ C}$ for gallium at $H = 18.125$ kG.

12. *The temperature dependence of* $\Delta R/R_{0^\circ C}$ *for* $H = 18.125$ *kG and* $H = 22.30$ *kG, with the field parallel to the* P_2*- and* P_3*-axes respectively.* The conjugated values of temperature and field strength for which gallium behaves tetragonally (established by the intersection of the field dependency curves of $\Delta R/R_{0^\circ C}$ in the extreme positions at a given temperature) may also be determined by the intersection of the extreme temperature dependency curves for a given field. These curves are drawn in the figs. 23 ($H = 22.30$ kG) and 24 ($H = 18.125$ kG), and give rise to the following remarks.

a. The values found in this way for the conjugated temperature and field strength are in good agreement with those previously quoted (69°K and 22.30 kG; 63°K and 18.125 kG).

b. With the field parallel to the P_2-axis ($a = \pm 90°$) the temperature dependence of $\Delta R/R_{0°C}$ is quite normal, *i.e.* $\Delta R/R_{0°C}$ increasing when the temperature is decreasing. This effect is more pronounced in the liquid hydrogen (10°K to 20°K) than in the liquid nitrogen range (50°K to 77°K).

c. With the field parallel to the P_3-axis ($a = 0°$), however, an anomalous course is found. Between 30°K and 80°K this curve shows an almost horizontal part, thus tending towards the normal curve and finally intersecting it.

d. In particular we notice that with the field parallel to the P_3-axis (and therefore perpendicular to the P_2-axis) an anomalous behaviour is found both for the field dependence and for the temperature dependence of $\Delta R/R_{0°C}$.

e. Our former idea that it is unlikely that a temperature will be found near 175°K where equal values of the resistance occur on extrapolating to zero field, is supported by the total absence of any indication of an intersection of the temperature dependency curves for the two extreme positions.

REFERENCES

1) W. J. d e H a a s and J. W. B l o m, Commun. Kamerlingh Onnes Lab., Leiden No. 229*b*; Physica, 's-Grav. **1**, 134, 1933–1934.
2) W. J. d e H a a s and J. W. B l o m, Commun. No. 237*d*; Physica, 's-Grav. **2**, 952, 1935.
3) W. J. d e H a a s and J. W. B l o m, Commun. No. 231*b*; Physica, 's-Grav. **1**, 465, 1933–1934.
4) M. K o h l e r, Ann. Physik (5) **32**, 211, 1938.
5) L. S c h u b n i k o w and W. J. d e H a a s, Commun. No. 210*a*; Proc. kon. Akad., Amsterdam **33**, 418, 1930.
6) E. J u s t i and H. S c h e f f e r s, Phys. Z. **37**, 383 and 475, 1936.
7) P. K a p i t z a, Proc. roy. Soc., London A **123**, 292, 1929.
8) W. J. d e H a a s, J. W. B l o m and L. S c h u b n i k o w, Commun. No. 237*b*; Physica, 's-Grav. **2**, 907, 1935.
9) W. J. d e H a a s and P. M. v a n A l p h e n, Commun. No. 225*a*; Proc. kon. Akad., Amsterdam **36**, 253, 1933.

CHAPTER III

RESULTS OF THE MEASUREMENTS
WITH P_2-, P_3- AND $P_{2,3}$-CRYSTALS, WHEN THE FIELD IS PERPENDICULAR TO THE CURRENT

Summary

For crystals with the current parallel to the shorter crystal axes (P_2- and P_3-axes) the transverse magneto-resistance effect shows a maximum when the field is parallel to the longer P_1-axis. The interchanging of these shorter axes is found again here, since the values of $\Delta R/R_{0°C}$ in the maxima are higher for the P_3-crystal than for the P_2-crystal at liquid nitrogen temperatures, whereas in the liquid hydrogen range the opposite is found. The temperatures where equivalence is found (depending on the field strength) seem to be lower than those for the P_1-crystal. Probably this is caused by a slight difference in purity of the crystals (which are necessarily different).

However, in the minima the interchanging is not found, for the values of $\Delta R/R_{0°C}$ for the P_3-crystal are always higher than those for the P_2-crystal.

For a crystal with the current parallel to the bisector of the angle between the P_2- and P_3-axes ($P_{2,3}$-crystal) values of $\Delta R/R_{0°C}$ are measured which lie, in general, between those for the P_2- and P_3-crystals.

1. *Introduction*. The measurements of the transverse effect have been continued with P_2-, P_3- and $P_{2,3}$-crystals [1]) (by a $P_{2,3}$-crystal we mean a single-crystal with the length parallel to the bisector of the angle between the P_2- and P_3-axes).

For each of these crystals rotational diagrams have been measured at five different temperatures; at each temperature this was done for three field strengths. The comparison of the results is made easier by giving the rotational diagrams, measured at a certain temperature, side by side for the three different crystals and drawn to the same scale.

Measurements can only be carried out here with different crystals, which have of course different physical and chemical impurities. So the value of $\Delta R/R_{0°C}$ may be greatly influenced by these variations.

Therefore special care must be taken here over the comparison of the results, and conclusions must be drawn with more reserve than with the P_1-crystals. The fact that the results found with different crystals may be compared at all is proved by the reproducibility of the measurements with various crystals of the same orientation. For all the crystals (P_1- as well as P_2-, P_3- and $P_{2,3}$) the mutual differences are only of the order of a few per cent for each type. This may be due to preparing all the crystals from the same material, so that the chemical impurity is fairly constant. In addition the assumption previously mentioned (see chapter I) that this causes only a slight disturbance in the lattice seems quite certain, for the residual resistance is quite small, and is even less than that for the purest of S c h u b n i k o w's bismuth [2]). This small residual resistance proves that the physical impurities are extremely low too. This might be due to storing the crystals at room temperature where they are just under the melting point. In effect, therefore, they are being annealed for some time before the beginning of the measurements and this is a well known method for diminishing the perturbations in the lattice. It may be assumed that in this way the most important irregularities have vanished in the period between the preparation of the crystal and the first experiment. We will refer to this point again in the next chapter.

The preceding discussion has shown that we are probably justified in comparing the results of different crystals even when they have different orientations, assuming that here again the impurities are always of the same order.

Some idea of the differences in impurity for the P_1-, P_2- and P_3-crystals which we used for our measurements may be deduced from the resistances determined near 14°K when establishing the temperature dependence of the ideal resistance. Since the temperature independent part of the resistance is already by far the most important in this region, the purest crystal (with the lowest residual resistance) must show the smallest value of $R/R_{0°C}$. The following values of $R/R_{0°C}$ were found [1]):

> 0.00276 for the P_1-crystal at 14.398°K,
> 0.00240 for the P_2-crystal at 14.334°K,
> 0.00194 for the P_3-crystal at 14.286°K.

The small temperature differences do not seriously affect the comparison. Hence the P$_3$-crystal may be considered the purest of the three, the P$_2$-crystal being the next and the P$_1$-crystal the impurest. In order to be able to refer when necessary to these conclusions we published for the P$_2$- and P$_3$-crystals only those measurements made with the specimens which were also used for the determination of the temperature dependence of the ideal resistance. For the P$_{2,3}$-crystals this dependence has never been measured.

In chapter VII it will be shown how the introduction of the Kohler diagram [3]) has removed all our troubles in this respect.

2. *The rotational diagrams at liquid nitrogen temperatures*. We decided to begin the discussion with the liquid nitrogen results here because we were especially anxious to see whether in this range the phenomena connected with the interchanging of the shorter axes, which were found with the P$_1$-crystals, will appear again. The diagrams we give together in figs. 1 ($T = 77.4°$K) and 2 ($T = 56.5°$K).

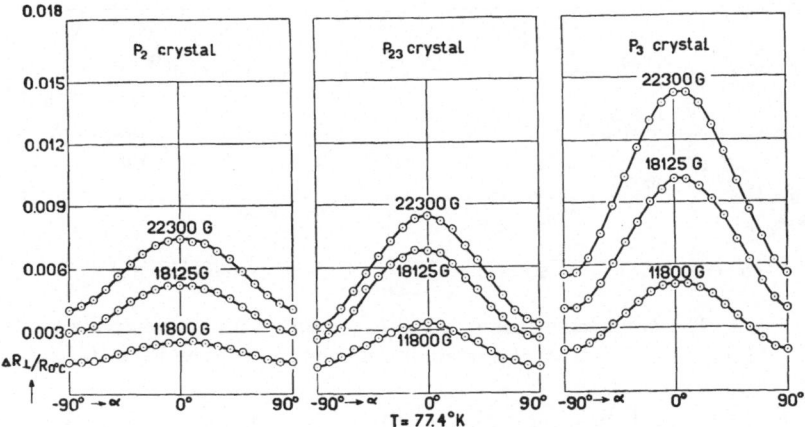

Fig. 1. The transverse magneto-resistance effect for single-crystals of gallium, as a function of the angle between the field and the P$_1$-axis.

Here a is the angle between the field and the P$_1$-axis for all three crystals. This was preferred because the P$_1$-axis was the only direction which was perpendicular to the current for all three. The following remarks can be made:

a. The interchanging of the shorter axes which was so very evident in the measurements with the P$_1$-crystal, so that all the results were influenced by it, does not appear for these crystals in this temperature range. The values of $\Delta R/R_{0°C}$ are always higher here for the P_3-crystal than those for the P_2-crystal.

It will be shown later in the discussion of the liquid hydrogen results that this happens to be one of the few cases where the difference in purity of the various crystals is important. By chance, the difference in resistance due to the impurity, although itself pretty small, is able to conceal the real state of affairs. This is because the values of $\Delta R/R_{0°C}$ for the P_2- and P_3-crystals are of the same magnitude in this temperature region, as might be expected from the results with the P_1-crystals. The P_3-crystal is the purest and consequently it shows values of $\Delta R/R_{0°C}$ which are higher than those for the P_2-crystal.

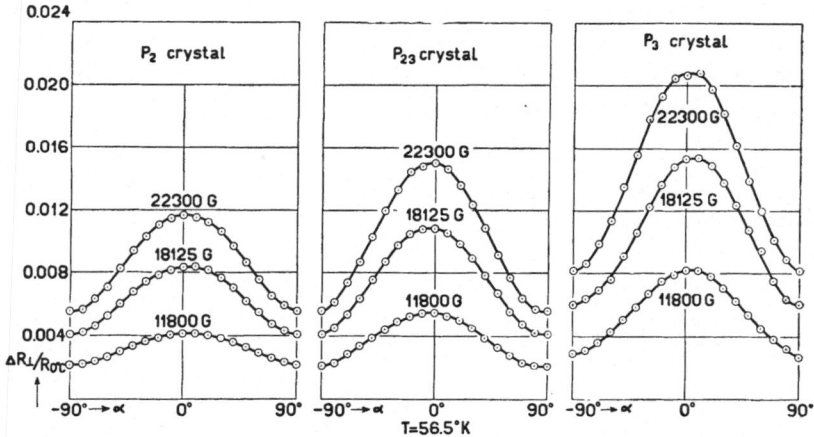

Fig. 2. The transverse magneto-resistance effect for single-crystals of gallium, as a function of the angle between the field and the P_1-axis.

b. In general, however, the results of the different crystals compare satisfactorily, in spite of slight differences in impurity. This, for example, may be seen from the fact that the values of $\Delta R/R_{0°C}$ for the $P_{2,3}$-crystal are usually between those found for the P_2- and P_3-crystals under similar conditions. At 56.5°K some exceptions are found where the differences are only of the order of the experimental error. At 77.4°K, however, some exceptional points are found (near the minimum) where the differences are greater. Here the $P_{2,3}$-crystal always shows a value of $\Delta R/R_{0°C}$ which is rather too small. This might easily be explained by assuming some stronger perturbations in the lattice for the $P_{2,3}$-crystal than for the other ones.

c. The rotational diagrams found here are always fully sinusoidal with a period of 180°.

d. As always a greater value of $\Delta R/R_{0°C}$ is found either when the field is increased or when the temperature is lowered.

e. The maximum is always found at $\alpha = 0°$, *i.e.* when the field is parallel to the P_1-axis. So for the minimum the field is parallel to one of the shorter axes or to their bisector. Since with the P_1-crystal the field is parallel to one of the shorter axes either when a minimum or when a maximum is measured, it may be expected that with these new crystals there will be a much greater ratio between the values of $\Delta R/R_{0°C}$ at $\alpha = 0°$ and at $\alpha \pm 90°$ than with the P_1-crystal under similar conditions. Between the P_1-axis and the other two there is a much greater anisotropy than between these shorter ones themselves which leads to this conclusion (since it is known that even a slight anisotropy strongly influences the value of $\Delta R/R_{0°C}$).

Indeed we found for the ratio in this case values between 1.7 and 2.6 at 77.4°K and between 1.9 and 2.8 at 56.5°K. With the P_1-crystals the value was appreciably smaller being about 1.1.

f. It is noticeable that in the maxima the values of $\Delta R/R_{0°C}$ are almost equal to those found in the maxima with the P_1-crystals under similar conditions, whereas in the minima they are less than a half of those of the P_1-case (the values obtained at *e*. also indicate a factor 2).

This cannot be caused by a difference in impurity of the crystals used, since we know the P_1-crystal to be the impurest, and so higher values would have been expected for the P_2- and P_3-crystals from this consideration.

3. *The rotational diagrams at* 20.4°K. The diagrams are found in fig. 3. On examination of the curves and after comparison with the liquid nitrogen results it is clear that in the range between these temperatures and the boiling point of hydrogen the interchanging of the shorter axes has occurred.

a. Firstly it may be observed from lowering the temperature. In this temperature region we find a higher value of $\Delta R/R_{0°C}$ at the maximum for the P_2-crystal than for the P_3-crystal at both the higher fields (22.30 kG and 18.125 kG). But at liquid nitrogen temperatures the opposite behaviour was found. Thus the temperatures at which the maxima have equal values, and where the shorter axes may be assumed to be equivalent, lie between 56.5°K and 20.4°K.

b. Secondly the interchanging of the P_2- and P_3-axes may be observed under the influence of increasing field strength just as with the P_1-crystals. For at 11.80 kG the maximum shows a somewhat lower value of $\Delta R/R_{0°C}$ for the P_2-crystal than for the P_3-crystal which is just the opposite behaviour to that at the higher fields. Hence with a field of 11.80 kG the situation is the same here as it was at 56.5°K, and it follows therefore that the interchanging of the shorter axes must take place below 20.4°K (because of the small difference it is probably only a little below).

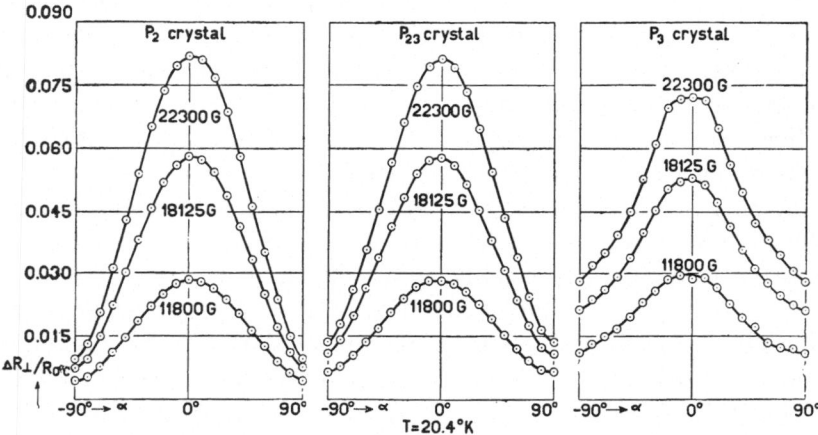

Fig. 3. The transverse magneto-resistance effect for single-crystals of gallium, as a function of the angle between the field and the P_1-axis.

We estimated this temperature (not only for 11.80 kG, but also for the higher fields) from the intersection of the temperature dependency curves of the maximum of $\Delta R/R_{0°C}$ for both the P_2- and P_3-crystals. At 11.80 kG this gave 20°K (*i.e.*, as predicted, a value slightly below 20.4°K), at 18.125 kG nearly 26°K and at 22.30 kG about 30°K. The latter temperatures are only approximate because of the absence of measuring points near the intersection which occurs in the range 50°K—20°K where measurements are difficult to make.

These approximate values of the temperatures at which the shorter axes are equivalent are found to be somewhat lower than those for the P_1-crystals in the same fields. In the same way therefore the temperature found on extrapolating the relation between temperature and field at which equivalence exists to zero field (about 10°K) is also lower here.

For the following reasons we think that our former data (30°K from the P_1-crystal measurements, 25°K from the "ideal" resistance curves) are much more reliable than this new value:

a. the temperature independent disturbances in the lattice were constant in the P_1-crystal measurements, since the same specimen was used all the time,

β. the influence of the temperature independent disturbances was eliminated in the ideal resistance method by subtracting the residual resistance,

γ. for the determination of this temperature with the P_2- and P_3-crystals only a few measurements were available.

Allowing for these considerations, and for the fact that different crystals had to be used, the agreement is quite satisfactory.

c. A possible explanation (previously mentioned when discussing the liquid nitrogen results) of this temperature being lower than the more reliable one found with the P_1-crystal, might be based on the fact that the P_3-crystal is purer than the P_2-crystal. If one assumes at liquid nitrogen temperatures, where $\Delta R/R_{0°C}$ is greater for the P_3-crystal, that the shorter axes are really equivalent, then the higher purity of this crystal will cause the intersection of the temperature dependency curves of $\Delta R/R_{0°C}$ to move to a lower temperature. Hence in the liquid nitrogen range the intersection should vanish and reappear in the liquid hydrogen region. The P_3-crystal might be assumed therefore to have really the smaller value in the latter range, but this is hidden by the effect of the higher purity. From this argument it might be expected that on lowering the temperature further the influence of the difference in purity will be less important, and may become a second order effect compared with the influence of the non-equivalence of the shorter axes. The much smaller values of $\Delta R/R_{0°C}$ shown by the P_3-crystal at the lower liquid hydrogen temperatures seem to agree with this conclusion.

In any case the important and fundamental effect of the interchanging of the axes is again seen.

d. It is seen that the interchanging of the shorter axes only appears for the maxima. For the minima we find that after lowering the temperature from 56.5°K to 20.4°K $\Delta R/R_{0°C}$ has a still higher value for the P_3- than for the P_2-crystal. However, two arrangements are concerned, which are crossed over by an interchange of the shorter axes. For, with the P_2-crystal the current is parallel to the P_2-axis

4

and in the minimum the field parallel to the P_3-axis; with the P_3-crystal it is just the opposite. At still lower temperatures this behaviour remains the same.

e. The deviations from the sinusoidal form of the rotational diagrams for the P_3-crystal will be discussed later with the lower temperature results.

f. With a single exception (which is of the same order as the experimental error) the values of $\Delta R/R_{0^\circ C}$ for the $P_{2,3}$-crystal lie between those for the other two crystals when under similar conditions.

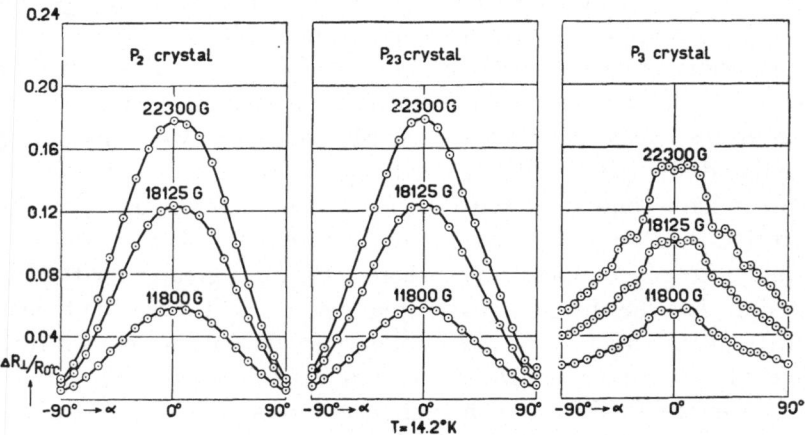

Fig. 4. The transverse magneto-resistance effect for single-crystals of gallium, as a function of the angle between the field and the P_1-axis.

4. *The rotational diagrams at* 14.2°K *and* 10.4°K. These diagrams are given in figs. 4 ($T = 14.2$°K) and 5 ($T = 10.4$°K). In general the results are in agreement with those found at 20.4°K and therefore only the new aspects which appear will be discussed.

a. In this case at 11.80 kG the maximum also shows a smaller value of $\Delta R/R_{0^\circ C}$ for the P_3-crystal than for the P_2-crystal, as might be expected from the measurements at 20.4°K.

b. The values of $\Delta R/R_{0^\circ C}$ for the $P_{2,3}$-crystal at 10.4°K do not lie so well between those for the other two as they did at the other temperatures. As a difference in purity should also have been perceptible at the other temperatures there must have been a special reason for this anomaly here. We think it is probably caused by the fact that 10.4°K is a solid hydrogen temperature and is much less reproducible than a liquid hydrogen temperature, even though the vapour pressure (measured with a manometer) is adjusted with great care to a given value. Since a temperature difference in this range has a great influence on the value of $\Delta R/R_{0°C}$ it might therefore be expected that the results will agree less well here. The experimental differences are compatible with the assumption that the temperature has been somewhat higher for the P$_2$-crystal measurements than for those with the other two. In support of this is the fact that at 10.4°K the values of $\Delta R/R_{0°C}$ for the P$_2$-crystal are not so much higher than those for the P$_3$-crystal as may be expected from the measurements at 20.4°K and 14.2°K.

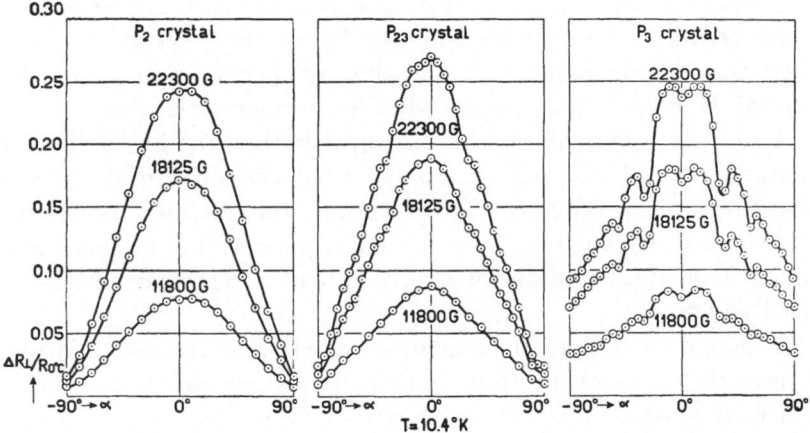

Fig. 5. The transverse magneto-resistance effect for single-crystals of gallium, as a function of the angle between the field and the P$_1$-axis.

c. For the P$_3$-crystal at these temperatures anomalies occur in the rotational diagrams at all field strengths (there was previously an indication of this at 20.4°K with the flattening of the maximum at 22.30 kG). At 10.4°K these anomalies are more extensive than at 14.2°K. In character they bear more resemblance to those found for bismuth than to those found for the P$_1$-crystals of gallium. For, firstly there are several secondary maxima on each side of the minimum which replaces the original maximum, instead of only one, and secondly their positions are independent of temperature and field strength.

Special attention may be drawn to the fact that the minimum remains undisturbed.

Since it is known from previous experience that the appearance of anomalies in the diagrams is stimulated by a higher purity, it is just possible that they exist only with the P_3-crystal (with the P_2-crystal there is no sign of them) because of its greater purity. However, after considering the comparability in other respects it seems that this difference is relatively unimportant, so we think that these singularities are characteristic of the P_3-crystal and not wholly due to its purity.

d. With the $P_{2,3}$-crystal at 10.4°K there are some indications of irregularities in the diagrams (most striking at 22.30 kG) and this supports our idea, since the $P_{2,3}$-crystal might be expected to behave in a manner intermediate between the other two crystals. The temperature at which the anomalies appear for the $P_{2,3}$-crystal may be a little too low on account of the slightly higher impurity of this crystal. However, this does not alter our former conclusion.

A second argument for our assumption that the anomalies are characteristic of the P_3-crystal, may perhaps be found from the fact that they appear only near the maximum, the minimum remaining undisturbed. Now it may be particularly noticed that at this maximum the field is simultaneously perpendicular to the current and to the P_2-axis.

Although for the present it is impossible to find an exact relation, it nevertheless seems profitable to recall here our former conclusion which, it must be remembered, was especially concerned with the measurements from a P_1-crystal (see chapter II).

"With the field parallel to the P_3-axis (and therefore perpendicular to the P_2-axis) an anomalous behaviour is found both for the field dependence and for the temperature dependence of $\Delta R/R_{0°C}$".

Since in the first chapters only the transverse effect is discussed, the above rule applies equally well to the situation where the field is simultaneously perpendicular to the current and to the P_2-axis.

REFERENCES

1) W. J. d e H a a s and J. W. B l o m, Commun. Kamerlingh Onnes Lab., Leiden No. 249c; Physica, 's-Grav. **4**, 767, 1937.
2) L. S c h u b n i k o w and W. J. d e H a a s, Commun. No. 207c; Proc. kon. Akad., Amsterdam **33**, 350, 1930.
3) M. K o h l e r, Ann. Physik (5) **32**, 211, 1938.

CHAPTER IV

RESULTS OF THE MEASUREMENTS WITH P_1-CRYSTALS, WHEN THE ANGLE BETWEEN THE FIELD AND THE CURRENT IS VARIED

Summary

In general, at liquid nitrogen temperatures the influence of β (the angle between the field and the current) on the magnetic increase of the resistance is normal for a gallium crystal with the length parallel to the longer crystal axis. Anomalies are only seen at the lower temperatures in this range. They become even more prominent at liquid hydrogen temperatures. These anomalies are probably due to a quite singular behaviour of the field dependence of the parallel effect at low temperatures. Here at the higher fields a tendency towards saturation is found.

The influence of the interchanging of the shorter axes, established in earlier experiments, agrees with our expectations.

There is an appreciable diminution of the impurity of the crystal, due to a long annealing process, reported.

1. *Introduction.* When varying the angle β between the field and the current the increase of the resistance, ΔR, for most metals shows a maximum value $\Delta R_\perp$, with the field perpendicular, and a minimum $\Delta R_{||}$, with the field parallel to the current. In general ΔR_β, the value corresponding to an arbitrary angle β ,may be calculated from [1]:

$$\Delta R_\beta = \Delta R_{||} \cos^2 \beta + \Delta R_\perp \sin^2 \beta. \tag{A}$$

Usually $\Delta R_{||}$ and $\Delta R_\perp$ are of the same order of magnitudo, $\Delta R_\perp / \Delta R_{||}$ usually lying between 1.5 and 2.5.

It was decided to investigate the behaviour of ΔR_β for gallium because we were very interested to see whether:

1) the formula (A) is also valid for this metal;
2) $\Delta R_\perp \approx 2\Delta R_{||}$ in this case also.

In addition the influences of temperature and field strength on ΔR_β have been investigated [2].

Firstly it may be reported that, independent of temperature and field strength, a minimum value of ΔR is indeed found when the field is parallel to the current. Using this result we could adjust the position of the support of the crystal to be exactly horizontal. The position of the support was changed slightly from that where it was judged to be best, and if only increases in ΔR occurred then the original position was known to be correct; otherwise suitable adjustments were made. To make the test more sensitive it was made at low temperatures (usually liquid hydrogen) and at the maximum available field strength.

The validity of formula (A) was tested by means of the rotational diagrams which should be sinusoidal in that case. We confined the research to the simplest cases where the field is turning about an axis which coincides with one of the crystal axes, thus being perpendicular to the current. For the P_1-crystal these are of course the P_2- and P_3-axes. As the electromagnet was rotating about the vertical, we carefully arranged either the P_2-axis or the P_3-axis to be vertical after adjusting the position of the support. For this reason it was very useful to have the shorter axes perpendicular to the side faces of the crystals.

In addition the exact position of the axis around which the field was rotating could also be examined. When for example the P_3-axis was supposed to be vertical, the field (which always remains horizontal when turned) was arranged to be perpendicular to the length of the crystal ($\beta = \pm 90°$). Since the field was now simultaneously perpendicular to the current, one of the previously measured values of $\Delta R/R_{0°C}$ should be reproduced, for this case coincides with the transverse effect when the field is parallel to the P_2-axis ($\alpha = \pm 90°$). Examining in this way the vertical position of the P_2-axis the earlier value for $\alpha = 0°$ should reappear (see chapter II).

At liquid nitrogen temperatures the comparison of the earlier values in the extreme positions with the new values of $\Delta R/R_{0°C}$ at $\beta = \pm 90°$ shows a reasonable agreement. Small differences (some per cent) which are found might be due to small temperature differences between the separate measurements. This is supported by the fact that nearly equal numbers of new values are too high and too low compared with the earlier ones. Since with the former liquid nitrogen measurements a minimum is found both for $\alpha = 0°$ and for $\alpha = \pm 90°$, small orientation inaccuracies would only have produced too high values.

At liquid hydrogen temperatures, however, there is quite a different situation. Here the following points may be noticed:

    a. the new values of $\Delta R/R_{0^\circ C}$ are always higher than the earlier ones,

    b. the differences increase with decreasing temperature,

    c. the differences are too large to be due simply to a temperature disagreement.

So here the discrepancies must have an other origin. They could be explained by assuming a decrease of the impurity in the crystals during the time which elapsed between the old and the new measurements. But then, however, the residual resistance should also have diminished in this time, and so also should the zero field resistance because of the relatively small temperature dependent part of the resistance in the liquid hydrogen region.

This dimunition was experimentally established and in addition the difference between the old values and these new ones proved to be larger with decreasing temperature, so the assumption seems quite certain. This effect was probably the result of the annealing process to which the P_1-crystal was submitted during the time (more than a year) between the first and the second series of measurements. In this time it was kept at room temperature which is only a few degrees below the melting point (30.2°C).

In the previous chapter we concluded that the most important disturbances in the lattice should have vanished during the annealing process which takes place in the time between the preparation of the single-crystal and the beginning of the measurements. To this can now be added the conclusion that further annealing for a long time diminishes the remaining irregularities still more.

After establishing this improvement in purity our first question was, if it should perhaps have caused the appearance of anomalies in the rotational diagrams for the transverse effect. The best chance of finding it would be at liquid hydrogen temperatures. So we decided to repeat in this region the results discussed in chapter II where necessary. However, the form of the diagrams is unchanged without any doubt.

We have used the results found with the field parallel to either one of the shorter axes for the comparison with the values found at $\beta = \pm 90°$ in the new measurements. The agreement is satisfactory now and confirms the conclusion drawn from the investigation at liquid nitrogen temperatures that the vertical position of the short axis around which the field was turning was sufficiently accurate.

Finally it may be noticed that a further test is possible, as the minima in the diagrams found when turning the field either round the P_2-axis or round the P_3-axis must show equal values of $\Delta R/R_{0°C}$. Of course in both cases $\Delta R_{||}/R_{0°C}$ is measured.

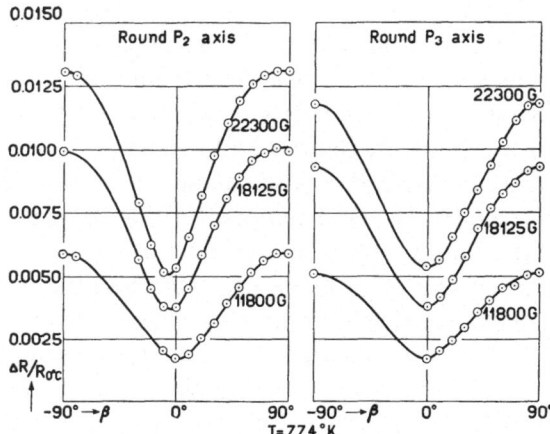

Fig. 1. The magnetic increase of the resistance as a function of the angle between the field and the current for a P_1-crystal.

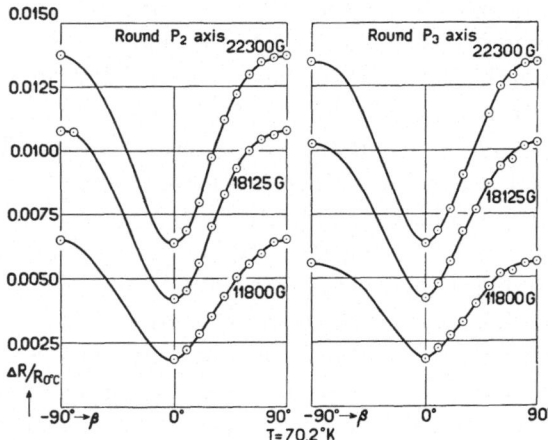

Fig. 2. The magnetic increase of the resistance as a function of the angle between the field and the current for a P_1-crystal.

2. *The rotational diagrams at liquid nitrogen temperatures.* We will begin the discussion with the results in the liquid nitrogen range, as

here again the influence of the interchanging of the shorter axes can be expected. For ease of comparison the diagrams with the field turning either round the P_2-axis or round the P_3-axis have been drawn to the same scale and are given side by side. They are given in the figs. 1 $(T = 77.4°K)$, 2 $(T = 70.2°K)$, 3 $(T = 56.5°K)$ and 4 $(T = 49.8°K)$.

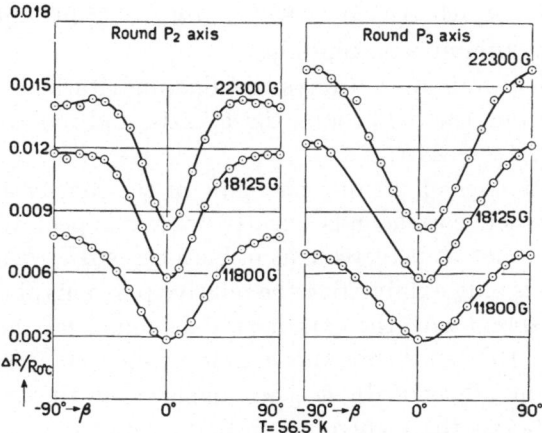

Fig. 3. The magnetic increase of the resistance as a function of the angle between the field and the current for a P_1-crystal.

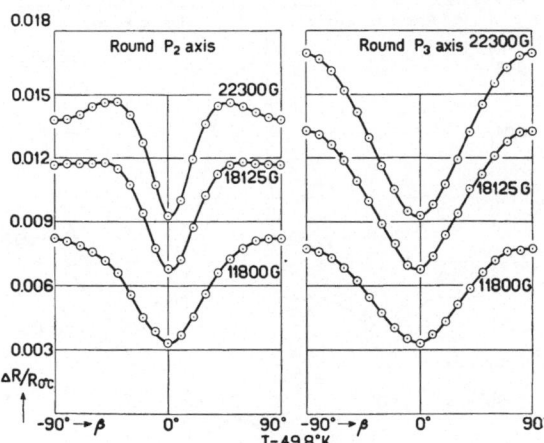

Fig. 4. The magnetic increase of the resistance as a function of the angle between the field and the current for a P_1-crystal.

The following remarks can be made:

a. When turning the field round the P_3-axis the rotational diagrams are sinusoidal; formula (A) is therefore valid.

When turning the field round the P_2-axis on the contrary it does not apply, although at 77.4 and 70.2°K the complications are only small (the minimum is somewhat sharper than the maximum). For the lower temperatures, however, they are important, especially at the stronger fields. For example it can be seen that at 22.30 kG the original maximum is replaced by a secondary minimum, at both 56.5 and 49.8°K, whereas at 11.80 kG and 18.125 kG only a flattening of the maximum is perceptible.

The original minimum always remains undisturbed.

b. On turning the field round the P_2-axis, and round the P_3-axis, it is found $\Delta R_\perp \approx 2\Delta R_\parallel$.

c. The influence of the interchanging of the shorter axes appears in a way which corresponds exactly to our expectations. In the minima it cannot be perceived, as in both cases $\Delta R_\parallel/R_{0°C}$ is measured, and so it is only exhibited in the relative positions of the maxima. These correspond with the extreme values found for the transverse effect and so the results are known beforehand. At 77.4 and 70.2°K the maximum values of $\Delta R/R_{0°C}$ are higher when the field is turned round the P_2-axis than when it is turned round the P_3-axis. At 56.5 and 49.8°K the relative values of the maxima depend on the field strength.

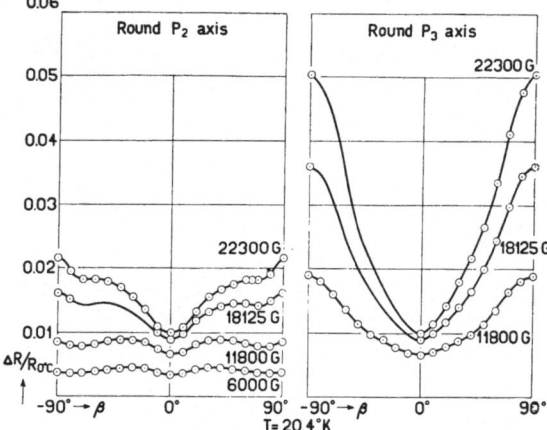

Fig. 5. The magnetic increase of the resistance as a function of the angle between the field and the current for a P_1-crystal.

3. *The rotational diagrams at liquid hydrogen temperatures.* The diagrams are drawn in the figs. 5 ($T = 20.4°K$), 6 and 7 ($T = 14.2°K$), 8 and 9 ($T = 10.4°K$). The results found at these temperatures

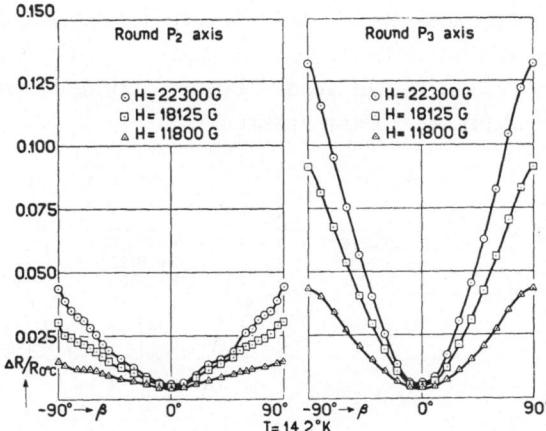

Fig. 6. The magnetic increase of the resistance as a function of the angle
between the field and the current for a P_1-crystal.

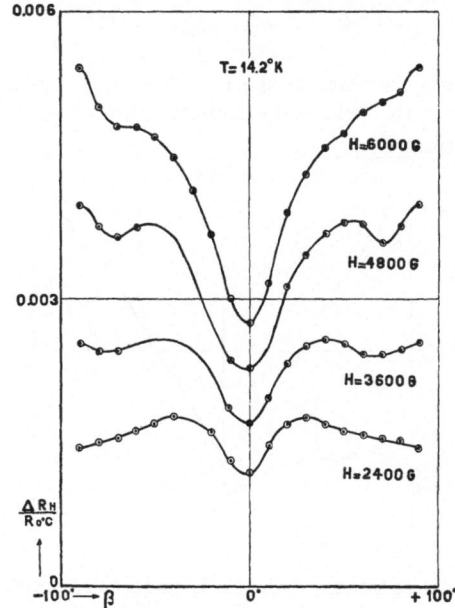

Fig. 7. The magnetic increase of the resistance as a function of the angle
between the current and the field, when this is turning round the P_2-axis.

made it necessary to measure some extra diagrams for the case
where the field was turning round the P_2-axis. These were at the
lower fields with a smaller magneto-resistance effect, so it was possi-

ble to use a larger scale in the two separate figures (7 and 9) to show
the singularities more effectively.

a. Firstly we discuss the results corresponding to the measurements at liquid nitrogen temperatures.

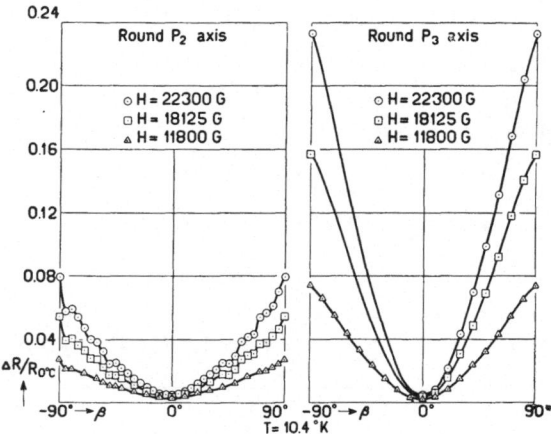

Fig. 8. The magnetic increase of the resistance as a function of the angle
between the field and the current for a P_1-crystal.

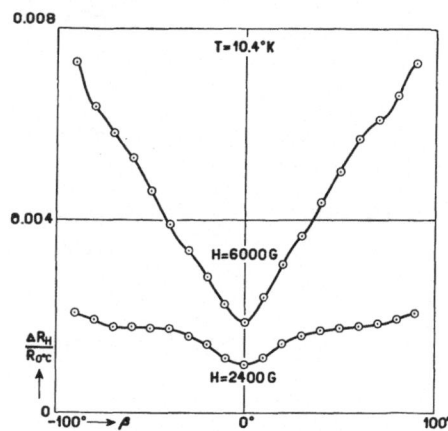

Fig. 9. The magnetic increase of the resistance as a function of the angle
between the current and the field, when this is turning round the P_2-axis.

a. As the interchanging of the shorter axes for all temperatures
and fields occurs above 30°K with this crystal, the maximum in the
diagrams when turning the field round the P_3-axis always shows a
higher value of $\Delta R/R_{0°C}$ than the maximum when turning round the
P_2-axis.

β. When turning the field round the P_3-axis the diagrams in this temperature range are still almost sinusoidal. Only for the higher fields are the maxima a little sharper than for the sine-curve.

b. Results unpredictable from those found in the liquid nitrogen region may be observed, however, when turning the field round the P_2-axis. At first correspondence seems probable, when only the diagram at 20.4°K and 6.00 kG is considered, for then, at $\beta = \pm 90°$ a secondary minimum is found which has about the same depth as the original one at $\beta = 0°$. As a matter of fact this result agrees quite well with the beginning of such a secondary minimum noticed at the lower liquid nitrogen temperatures, as, in general, secondary minima are intensified by decreasing temperature. Apart from this, the fact that the values of $\Delta R/R_{0°C}$ at $\beta = 0°$ and at $\beta = \pm 90°$ are nearly equal, so that the orientation of the field has practically lost its influence on $\Delta R/R_{0°C}$, is very unusual (for normally it is extremely important).

Although from previous experience it might be expected that either for stronger fields or at lower temperatures the size of the secondary minima should gradually increase, ending finally in an interchanging of minimum and maximum, the contrary was actually observed. For, firstly the secondary minimum at $\beta = \pm 90°$ has vanished at 20.4°K for the higher fields (11.80, 18.125 and 22.30 kG) and it has even been replaced by a new maximum. Secondly this new maximum is seen for the weaker fields at 14.2 and 10.4°K, as may be noticed by the comparison of the diagrams for 6.00 kG. We have tried to reestablish at the lower temperatures the situation where there is a secondary minimum with almost the depth of the original one (such as was found at 20.4°K and 6.00 kG), by measuring at lower fields. At 14.2°K we successfully attained this at 2.40 kG, while at some intermediate fields (3.60 and 4.80 kG) the new maximum was gradually appearing. At 10.4°K on the other hand no secondary minimum was perceptible even at the lowest field available (2.40 kG). As previously mentioned further lowering of the field is of no advantage because of the decreasing accuracy.

4. *The ratio $\Delta R_\perp/\Delta R_\parallel$ at liquid hydrogen temperatures.* This ratio which at liquid nitrogen temperatures is practically independent of temperature and field strength (showing the normal value of about 2) in the liquid hydrogen region varies considerably under different conditions.

With the field turning round the P_2-axis a too small value (approximately 1) is found at low fields (6.00 kG at 20.4°K and 2.40 kG at 14.2°K), and hence the value of $\Delta R/R_{0°C}$ is then almost independent of the orientation of the field as was mentioned previously. An exceedingly high value is, on the contrary, found at the strong fields, *i.e.* for 22.30 kG $\Delta R_\perp/\Delta R_\parallel$ has the normal value 2 at 20.4°K, while it is 7 at 14.2°K and even 17 at 10.4°K.

With the field turning round the P_3-axis a similar effect is found to a still higher degree. For 22.30 kG $\Delta R_\perp/\Delta R_\parallel$ is 5 at 20.4°K, 20 at 14.2°K and 50 at 10.4°K. Now it may be noticed that the value of $\Delta R_\perp$ in this case corresponds with the previously measured value of $\Delta R_\perp$ for $\alpha = \pm 90°$, since in both cases the transverse effect for a P_1-crystal with the field parallel to the P_2-axis is concerned. From these old measurements it is known that both the field and the temperature dependences of $\Delta R_\perp/R_{0°C}$ are quite normal with this direction of the field (see chapter II). So consequently the origin of the singular behaviour of the ratio $\Delta R_\perp/\Delta R_\parallel$ should be found in an abnormal course of $\Delta R_\parallel/R_{0°C}$.

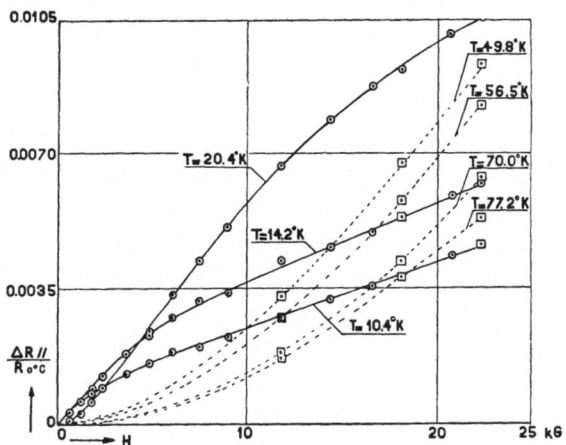

Fig. 10. Field dependence of the parallel effect for a P_1-crystal.

5. *The field dependence of $\Delta R_\parallel/R_{0°C}$.* In fig. 10 the field dependences of $\Delta R_\parallel/R_{0°C}$ at the liquid hydrogen temperatures are drawn as full lines, whereas those concerning liquid nitrogen temperatures are drawn as dotted lines. For the construction of these last lines only a few points were available (*i.e.* those measured at $\beta = 0°$ for the

determination of the rotational diagrams at these temperatures). Their construction was nevertheless possible, because of the normal course of the curves at liquid nitrogen temperatures. This means that they are fully parabolic. So, for the parallel effect, just as for the transverse effect, there is no trace of a linear part in the curve (see chapter II).

With regard to the results in the liquid hydrogen region it may be noticed:

a. When at 20.4°K only the course at low fields (up to 5 kG) is considered a normal behaviour seems very probable here too. Going to higher fields at 20.4°K a gradual change into a linear part is occurring. Now, although this is a new effect for gallium, it agrees with the normal behaviour which K a p i t z a [3]) established when he found the beginning of a linear part at lower fields when the temperature is decreased.

So it seems propable that for the parallel effect this change into a linear part does indeed exist, and that it begins at a field lower than the limit of our electromagnet.

When the curves are also considered only at low fields at 14.2 and 10.4°K, the absence of the parabolic beginning may be observed. This is compatible with the assumption that the intermediate region between the parabolic and linear parts of the curve, which normally moves to lower fields when the temperature is decreased, is concealed here below the minimum field where it is possible to measure with sufficient accuracy. So, since the parabolic part cannot be measured, there is a linear part straight away.

Thus far there are no definite anomalies to be seen.

b. At the higher fields on the contrary the curves become quite anomalous. Thus at 20.4°K the linear part is very short, and from 5 kG there follows a gradual decrease in the slope, which means a lessening influence of the field on $\Delta R_{||}/R_{0°C}$. Initially we thought that it might reach a saturation value, but this idea was found to be wrong when we considered the curves at the lower temperatures. It may be stated then that up to the highest fields available a certain influence of the field on $\Delta R_{||}/R_{0°C}$ persists. On closer examination a second linear part with a smaller slope seems more probable now. Thus, allowing for the imperceptible parabolic part, at 14.2 and at 10.4°K the field dependency curve of $\Delta R_{||}/R_{0°C}$ consists of two consecutive linear parts continuously connected. In addition the second linear parts prove to be parallel at both temperatures and it is not impossible that at 20.4°K the curve will finally be parallel also when continued to higher fields. Unfortunately we had no opportunity to investigate this because of the limit of our electromagnet.

c. It may be deduced from the above that at 20.4°K the beginning of the second linear part lies above the maximum field available (22.30 kG) and this can also be observed in the graphs. Since at 14.2°K it begins at nearly 10 kG and at 10.4°K at nearly 5 kG, the conclusion seems to be that this field strength decreases with the temperature. If we assume the behaviour of $\Delta R_{||}/R_{0°C}$ at still lower temperatures to be the regular continuation of that found here, we may presume that at liquid helium temperatures the first linear part will perhaps vanish as well. This would result in a single straight line through the origin, parallel to the second part of the curve found at 10.4°K. We regret we have missed the chance to verify this.

d. The decrease in slope found for the curves at the liquid hydrogen temperatures has the remarkable consequence that the dotted lines, giving the field dependence of $\Delta R_{||}/R_{0°C}$ at liquid nitrogen temperatures, are finally intersected by them. This means that for sufficiently high field strengths the value of $\Delta R_{||}/R_{0°C}$ is, contrary to the normal behaviour, less in the liquid hydrogen than in the liquid nitrogen region. In the liquid hydrogen range this decrease of $\Delta R_{||}/R_{0°C}$ with the temperature continues, as may be seen firstly from the intersection of the curve for 20.4°K by both the others, and secondly by the fact that the curve for 10.4°K lies totally below that for 14.2°K. The intersection for these last two must lie at a very weak field.

6. *The temperature dependence of* $\Delta R_{||}/R_{0°C}$. To illustrate the abnormal behaviour of the temperature dependence of $\Delta R_{||}/R_{0°C}$ it is plotted graphically for several field strengths in fig. 11.

a. It may be seen that at liquid nitrogen temperatures $\Delta R_{||}/R_{0°C}$ normally increases when the temperature is lowered, whereas in the liquid hydrogen range the opposite is found for all field strengths (*i.e.* not only for the stronger fields). Consequently a maximum value of $\Delta R_{||}/R_{0°C}$ must be found in the intermediate temperature range, and for all field strengths this is actually found between 20 and 30°K. On account of the absence of measurements in this region the different temperatures could only be estimated. From the results we can only report that the temperature where the maximum is found decreases with the field strength.

b. As will be later discussed in chapter VI, K o h l e r [4]) has made clear that physically the quotient $\Delta R/R_{0T}$ has a certain significance. Since from the Kohler diagram it follows that $\Delta R_{||}/R_{0T}$ does not decrease with the temperature it is clear that the decrease mentioned above in *a.* refers only to a pseudo-effect due to the use of the quotient $\Delta R/R_{0°C}$. Therefore no further attention will be given to it. It must be particularly noted that, although for $\Delta R_{||}/R_{0T}$ a decrease is absent, there is still an anomalous temperature dependence of this quantity, as was deduced from the enormous increase of the ratio $\Delta R_{\perp}/\Delta R_{||}$, where the influence of $R_{0°C}$ is eliminated (see section 4). With the Kohler diagram this will be considered in more detail.

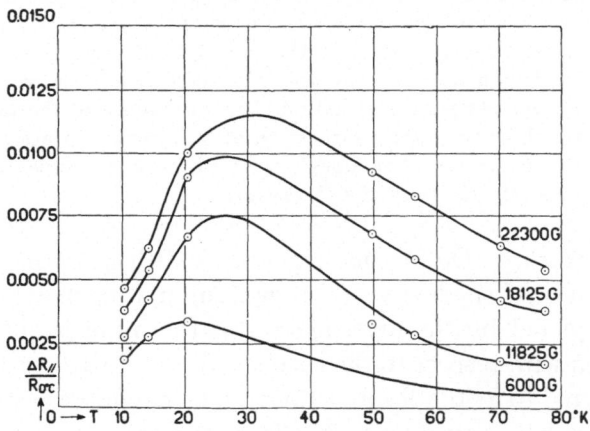

Fig. 11. Temperature dependence of the parallel effect for a P₁-crystal.

c. From the previously mentioned effect (found with the transverse field) that the magnetic increase becomes temperature independent in the same region as the normal resistance, we think we are justified in saying that in the liquid helium range the parallel effect will be constant too, although unfortunately there has been no experimental verification.

REFERENCES

1) W. J. d e H a a s and P. M. v a n A l p h e n, Commun. Kamerlingh Onnes Lab., Leiden No. 225*a*; Proc. kon. Akad., Amsterdam **36**, 253, 1933.
2) W. J. d e H a a s and J. W. B l o m, Commun. No. 249*d*; Physica, 's-Grav. **4**, 778, 1937.
3) P. K a p i t z a, Proc. roy. Soc., London A **123**, 292, 1929.
4) M. K o h l e r, Ann. Physik (5) **32**, 211, 1938.

CHAPTER V

RESULTS OF THE MEASUREMENTS WITH P_2- AND P_3-CRYSTALS, WHEN THE ANGLE BETWEEN THE FIELD AND THE CURRENT IS VARIED

Summary

For single-crystals of gallium, with the shorter crystal axes parallel to the current, the rotational diagrams giving the increase of the resistance ΔR_H in a magnetic field as a function of the orientation of the field with respect to the current, are simple sine-curves. As in the case of the P_1-crystals (with the long axis parallel to the current) $\Delta R_{||}$ shows, in contrast with the normal behaviour at liquid hydrogen temperatures, a diminishing influence of the field at high field strengths. This suggests a saturation at still higher values of the field. Also $\Delta R_{||}/R_{0°C}$ decreases anomalously with decreasing temperature in the same manner as it does with the P_1-crystals.

1. *Introduction.* Our earlier experiments [1] [2] [3] on the change in the resistance of single-crystals of gallium in a magnetic field have been completed by the investigation of the influence of the field orientation with respect to the current. We used two crystals, hereafter referred to as the P_2- and P_3-crystals, having respectively one of the shorter P_2- and P_3-axes parallel to the length (*i.e.* to the direction of the current). Previously we have shown for the P_1-crystal (where the long axis is parallel to the current) that ΔR_β (the increase of the resistance when the angle between field and current has an arbitrary value β) may be calculated from:

$$\Delta R_\beta = \Delta R_{||} \cos^2\beta + \Delta R_\perp \sin^2 \beta, \qquad (A)$$

where $\Delta R_{||}$ is the minimum value found when the field is parallel, and $\Delta R_\perp$ is the maximum value found when it is perpendicular to the current. We wished to see if this is also true for the P_2- and P_3- crystals.

Again we confined the measurements to the cases where the field is rotated round the two axes perpendicular to the length of the crystal [1]). Thus for the P_2-crystal, rotational diagrams (giving $\Delta R/R_{0°C}$ in dependence on β) have been established by turning the field either round the P_1-axis or round the P_3-axis; for the P_3-crystal the turning has been round the P_1- and the P_2-axes.

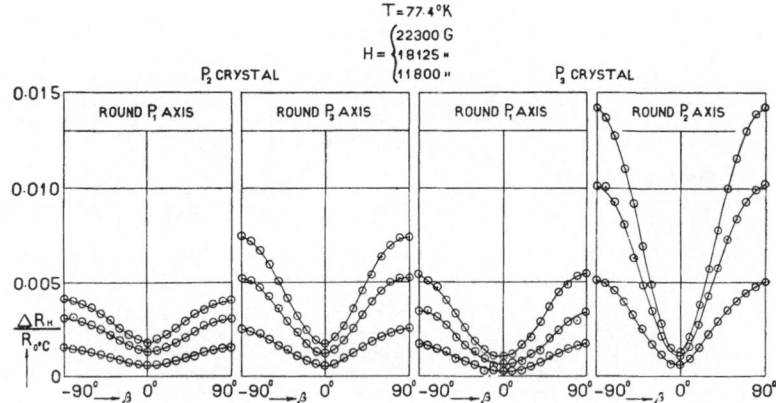

Fig. 1. The magnetic increase of the resistance as a function of the angle between the field and the current at 77.4°K.

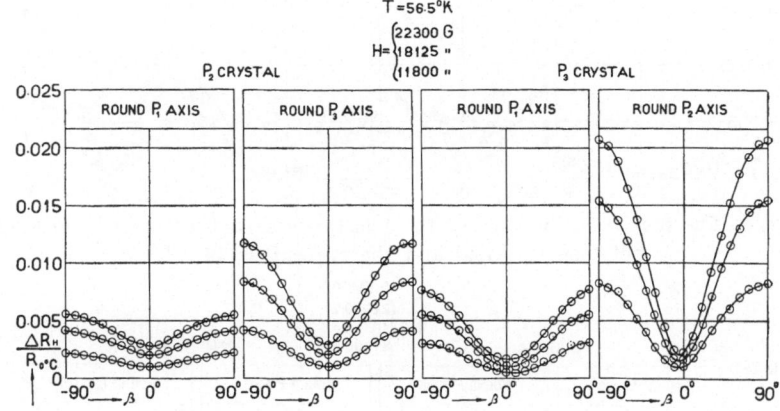

Fig. 2. The magnetic increase of the resistance as a function of the angle between the field and the current at 56.5°K.

2. *The rotational diagrams.* The rotational diagrams are drawn in the figs. 1 $(T = 77.4°K)$, 2 $(T = 56.5°K)$, 3 $(T = 20.4°K)$, 4 $(T = 14.2°K)$ and 5 $(T = 10.4°K)$. For easier comparison the four diagrams for the two crystals have been brought together into one figure side by side, on the same scale.

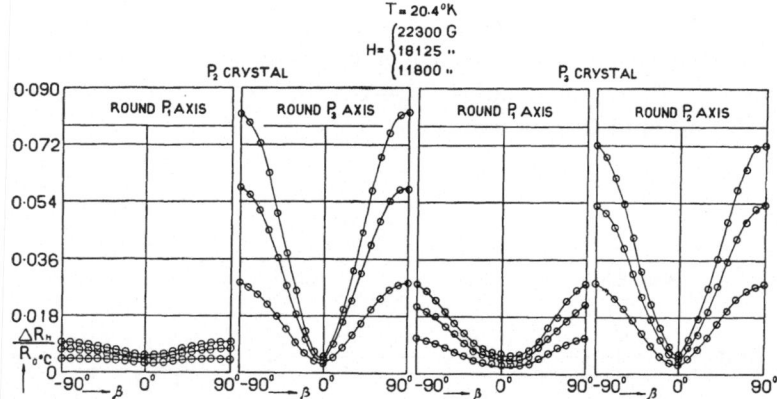

Fig. 3. The magnetic increase of the resistance as a function of the angle between the field and the current at 20.4°K.

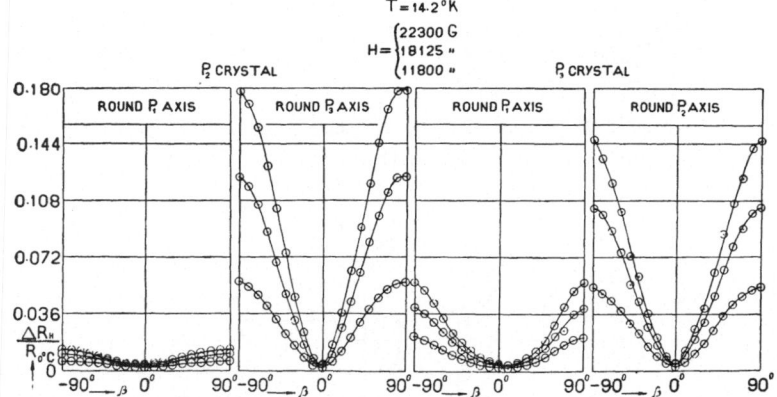

Fig. 4. The magnetic increase of the resistance as a function of the angle between the field and the current at 14.2°K.

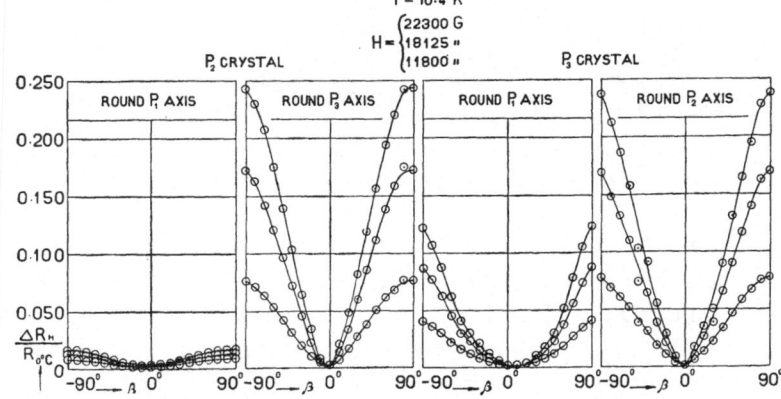

Fig. 5. The magnetic increase of the resistance as a function of the angle between the field and the current at 10.4°K.

The following details are accentuated.

a. A minimum is always found when the field is parallel ($\beta = 0°$) and a maximum when it is perpendicular to the current ($\beta = \pm 90°$) in accordance with the general rules.

b. The diagrams are always sinusoidal, which implies that the formula (A) can be applied unrestricted.

c. The maximum values agree, in general, with the results of the earlier measurements in which the field was turned in a plane perpendicular to the length [2]). Only at 10.4°K is the agreement bad. This is almost certainly due to the fact that this temperature (where the hydrogen is solid) is much less reproducible than the liquid hydrogen temperatures. As ΔR in this temperature range is strongly influenced by the temperature, small temperature differences will produce large discrepancies, compared with the old measurements.

3. *The ratio* $\Delta R_\perp / \Delta R_\parallel$. The ordinary value of the ratio $\Delta R_\perp / \Delta R_\parallel$ is about 2. At liquid nitrogen temperatures the only exception was observed for the P_3-crystal, when the field was turning round the P_2-axis. Then we found $\Delta R_\parallel$ much less than $\Delta R_\perp$.

At liquid hydrogen temperatures $\Delta R_\parallel$ is always much less than $\Delta R_\perp$, both for the P_2- and P_3-crystals, and as in the case of the P_1-crystals this is due to the abnormal behaviour of the field and the temperature dependences of $\Delta R_\parallel / R_{0°C}$.

4. *The field dependence of* $\Delta R_\parallel / R_{0°C}$. As with the P_1-crystals, the field dependence has been studied for the P_2- and P_3-crystals, espe-

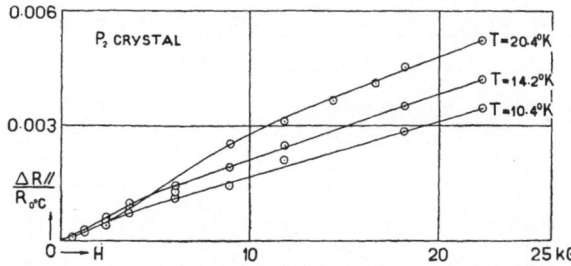

Fig. 6. The parallel effect as a function of the field strength for the P_2-crystal at liquid hydrogen temperatures.

cially at the liquid hydrogen temperatures. The first columns of tables I (P_2-crystal) and II (P_3-crystal) give the value of the field

TABLE I

H	$\Delta R_{||}/R_{0°C}$ for a P_2-crystal		
	$\Delta R/R_{0°C}$		
	$T = 20.4°K$	$T = 14.2°K$	$T = 10.4°K$
0.600	0.00009	0.00009	0.00007
1.200	0.00021	0.00021	0.00028
2.400	0.00048	0.00060	0.00041
3.600	0.00087	0.00094	0.00071
6.000	0.00126	0.00137	0.00110
8.950	0.00252	0.00189	0.00140
11.800	0.00311	0.00247	0.00210
14.400	0.00366		
16.600	0.00410		
18.125	0.00452	0.00349	0.00286
22.300	0.00519	0.00419	0.00347

TABLE II

H	$\Delta R_{||}/R_{0°C}$ for a P_3-crystal		
	$\Delta R/R_{0°C}$		
	$T = 20.4°K$	$T = 14.2°K$	$T = 10.4°K$
1.200	0.00019		
2.400	0.00038		
3.600	0.00057		
4.800	0.00095		
6.000	0.00133		
8.950	0.00243		
11.800	0.00264	0.00266	0.00118
14.400	0.00472		
16.600	0.00537		
18.125	0.00483	0.00401	0.00118
22.300	0.00618	0.00398	0.00151

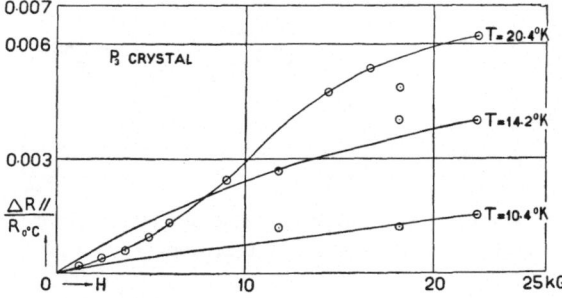

Fig. 7. The parallel effect as a function of the field strength for the P_3-crystal at liquid hydrogen temperatures.

strength, and columns 2, 3 and 4 give the values of $\Delta R_{||}/R_{0°C}$ at 20.4, 14.2 and 10.4°K. The curves are drawn in figs. 6 (P_2-crystal) and 7 (P_3-crystal). As in this temperature range the value of $\Delta R_{||}/R_{0°C}$ is very low, the inaccuracy of the measurements only permits an average curve to be drawn, and this is especially so for the P_3-crystal, where moreover very few measurements were available, so that the curve is very disputable. Nevertheless we are probably justified in saying that the results for these crystals are in agreement with those found earlier for the P_1-crystal [1]), where the following behaviour was noticed.

a. At 20.4°K the curve denoting the field dependence of $\Delta R_{||}/R_{0°C}$ consists of three parts:

1. a normal parabolic initial part,

2. a short linear part with a strong field dependency, following the parabolic one and similar to the curves K a p i t z a [4]) found for some other materials,

3. a second linear part of smaller slope than the first, which is quite unusual; this suggests a saturation at still higher values of the field, although this has not yet been found.

b. At 14.2 and 10.4°K the parabolic part, if it exists, is vanishing at the weaker fields and therefore we observe only the two linear parts.

c. Apparently the direction of the second linear part is independent of the temperature. Thus these parts are parallel for the different liquid hydrogen temperatures.

d. As the curve for 20.4°K is intersected by the curve for 14.2°K, at the higher field strengths the value of $\Delta R_{||}$ must be less at 14.2°K than at 20.4°K. This includes a deviation from the normal behaviour of the temperature dependence of $\Delta R_{||}/R_{0°C}$. The intersection of the two curves is due to the fact that the second linear part begins at a lower field strength when the temperature is decreased.

e. The intersection of the curve for 10.4°K with the others must occur at a very low field strength and is therefore imperceptible. Hence this curve is still lower than that for 14.2°K. Evidently the diminution of $\Delta R_{||}/R_{0°C}$ with falling temperature is continued in this temperature range.

5. *The temperature dependence of* $\Delta R_{||}/R_{0°C}$. *a.* Just as in the case of the P_1-crystals studied earlier, the value of $\Delta R_{||}/R_{0°C}$ must show

a maximum between liquid nitrogen and liquid hydrogen temperatures, for in the first temperature range we normally find an increase of $\Delta R_{||}/R_{0°C}$ with falling temperatures, whereas in the second range we find the opposite for both the P_2- and P_3-crystals. Since the temperature interval between liquid nitrogen and liquid hydrogen temperatures is rather inaccessible for experiments no measurements have been made there. The exact position of the maximum is therefore not known, although an estimation shows that it must lie between 20 and 30°K for both crystals. For the P_1-crystal the same limits have been found.

b. Comparing the values of $\Delta R_{||}/R_{0°C}$ for the P_2- and P_3-crystals we arrive at the conclusion that at liquid nitrogen temperatures the P_2-crystal always has a higher value than the P_3-crystal. At liquid hydrogen temperatures on the other hand the P_3-crystal shows, in general, the higher values. With a single exception, which may be a consequence of the inaccuracy of the measurements mentioned before, it seems reasonable to suppose that here again a phenomenon appears whereby the interchanging of the shorter axes [3] takes place in the range between liquid nitrogen and liquid hydrogen temperatures.

c. The values of $R/R_{0°C}$ for the two crystals at 14.3°K [2] (i.e. 0.00194 for the P_3- and 0.00240 for the P_2-crystal) indicate that the P_3-crystal is purer than the P_2-crystal, and one may at first think that this is the reason for the higher values of $\Delta R_{||}/R_{0°C}$ for the P_3-crystals at liquid hydrogen temperatures. However, this can not be so, for the higher purity would also produce greater values of $\Delta R_{||}/R_{0°C}$ at liquid nitrogen temperatures. In fact, in our opinion, the interchanging of the shorter crystal axes is once again clearly indicated. This proves once more the comparability of the results of the various crystals, as this behaviour was also found earlier in spite of using different specimens. Actually there were only slight differences in the purity, since the crystals were all grown in the same way from the same material [1].

d. The number of measurements and their relative inaccuracy were insufficient to enable us to investigate the relation between the temperature and the field strength, at which the shorter axes are equivalent. From earlier experiments one might perhaps expect that with decreasing temperature an increasing field strength will be found.

6. *Comparison with the P_1-crystal.* Comparing the results found for the P_2- and P_3-crystals with those found for the P_1-crystal, we have particularly noted that the values of $\Delta R_{||}/R_{0°C}$ for the various crystals are less divergent than might have been anticipated from previous experience. Moreover the temperature and field dependences bear a striking resemblance in both cases. It appears that the orientation of the crystal axes has not the important influence on the value and the course of $\Delta R_{||}/R_{0°C}$ as it has on the value and the course of $\Delta R_{\perp}/R_{0°C}$.

REFERENCES

1) W. J. d e H a a s and J. W. B l o m, Commun. Kamerlingh Onnes Lab., Leiden, No. 249d; Physica, 's-Grav. **4**, 778, 1937.
2) W. J. d e H a a s and J. W. B l o m, Commun. No. 249c; Physica, 's-Grav. **4**, 767, 1937.
3) W. J. d e H a a s and J. W. B l o m, Commun. No. 231b; Physica, 's-Grav. **1**, 465, 1933–1934.
4) P. K a p i t z a, Proc. roy. Soc., London A **123**, 292, 1929.

CHAPTER VI

THE KOHLER DIAGRAM FOR GALLIUM

Summary

On plotting the results of our measurements on the magnetic increase in resistance of single-crystals of gallium as Kohler diagrams, characteristic functions are found, both for the transverse and the parallel orientation of the field with respect to the current. This facilitates the observation of the results for different temperatures and also enables one to compare the measurements from a number of crystals. It is remarkable that for the transverse effect the trivalent Ga fits in with the bivalent Cd-type, introduced by J u s t i, and does not resemble the In-behaviour. For the parallel effect a saturation tendency exists, which is propably a general phenomenon.

1. *Introduction.* Since the publication of our experiments on the magnetic increase of the resistance of single-crystals of gallium K o h l e r [1]) has introduced a new graphical system for summarizing the results of the measurements. It is based on the experimental observation that the magnetic increase in resistance, ΔR, is becoming larger when the irregularities of the crystal lattice are diminishing. This is true whether the decrease in the irregularities has been caused by decreasing the temperature or by purifying the crystal. The intensity of these irregularities increases with $R_{0T}/R_{0°C}$ where R_{0T} and $R_{0°C}$ are the resistances in zero field, at $T°K$ and at $0°C$ respectively.

Now, in general the magnetic increase of the resistance increases with rising field strength, so this implies that if $\Delta R/R_{0T}$ has been diminished, in constant field, by an increase of $R_{0T}/R_{0°C}$, it may be restored to its initial value by an increase of the field strength.

Considering some of the results of J u s t i and his collaborators[2]), K o h l e r realised that the initial value of $\Delta R/R_{0T}$ could be restored by doubling the field strength H, when the irregularities in the lattice had been increased before in such a way that $R_{0T}/R_{0°C}$ is doubled. Obviously in both cases the ratio $H/(R_{0T}/R_{0°C})$ is the same.

Hence, assuming that equal values of $H/(R_{0T}/R_{0°C})$ always correspond to the same values of $\Delta R/R_{0T}$, K o h l e r suggested that
these two functions are plotted graphically. Then, if the assumption
is correct, a single curve should be obtained which is independent of
both the temperature and the purity of the crystal.

The functions are now plotted logarithmically (following a further
suggestion by K o h l e r) so that the exponent of the law connecting them may be deduced. In this form the graph is commonly
known as a "Kohler diagram".

When J u s t i c.s. [3]) first applied this method to their measurements the results were in general in good agreement with K o h-
l e r's assumption. They gave a single curve which can be considered
as a "characteristic function" for the metal in question, and this may
be regarded as a magnetic analogy to Matthiessen's rule [4]). Disagreements occur for both rules in similar cases and are particularly
noticeable for the non-cubic metals. For these metals we usually find
different results for crystals with different orientations. The situation with the Kohler diagram is even more complicated than is
the case with Matthiessen's rule, since here it is not only the
orientation of the current with respect to the crystal axes which is
important, but also the orientation of the field. When a polycrystal
is used, then an average curve between those valid for the separate
crystals may be expected. With J u s t i's [5]) measurements the
exact crystallographic conditions were not always known and hence
in the case of crystals of the non-cubic metals, it is not surprising
that the results showed a certain amount of spreading.

J u s t i c.s. [6]) established that even for the cubic metals the orientation of the field with respect to the crystal axes has a considerable
influence on the increase of the resistance at low temperatures. This
was determined by means of single-crystals with the current almost
parallel to one of the principle axes.

2. *The "reduced Kohler diagram"*. Following up his former proposal K o h l e r suggested an important refinement of the "Kohler
diagram" which enables us to compare the results found for the
different metals.

His fundamental idea is that for the purpose of comparing the
irregularities of the lattice for different crystals of the same material,
the value of $R_{0T}/R_{0°C}$ provides a proper scale (the division by $R_{0°C}$

occurs only to make the results independent of the size of the crystal).
However, for the comparison of the results of different metals he
pointed out that an arbitrary point on the temperature scale, such as
zero degrees centigrade, has no significance and instead one may
consider two different metals to be in equivalent conditions when
they are at the same Debije temperature(Θ). He therefore proposed
that $R_{0T}/R_{0\Theta}$ should be taken for the scale of the irregularities in the
lattice and that thus $\Delta R/R_{0T}$ should be plotted against $H/(R_{0T}/R_{0\Theta})$.
Such a graph is known as the "reduced Kohler diagram".

3. *The metallic types in the "reduced Kohler diagram"*. When
J u s t i *c.s.* plotted their results in the "reduced Kohler diagram" it
appeared that several metals could be collected into groups, which
each occupied a narrow strip in the diagram. The metals belonging
to the same group form a type and J u s t i calls each type by the
name of one of its member metals, *e.g.* Na-type, Al-type *etc.* The
characteristics of the various types are:
  a. the column of the periodic system to which the metals belong,
  b. the crystal structure into which they crystallize.
  There is one difficulty, in that for many of the metals the Debije temperature is only inaccurately known. J u s t i has usually
estimated an approximative value with the Lindemann formula
for the melting point:

$$\Theta = 137\sqrt{(T_s/MV^{2/3})}$$

(where T_s = melting point, M = atomic weight, V = atomic volume).
Hence it is possible that the part of the diagram occupied by a certain
type is much more narrow than has been indicated by J u s t i.
  J u s t i has hitherto distinguished between eight types, four of
them concerning oddvalency metals:
  1. Na-type [6]) (monovalent, cubic),
  2. Au-type [7]) (monovalent, cubic, face centered),
  3. Al-type [8]) (trivalent, cubic face centered),
  4. Ta-type [7]) (fivevalent),
and four concerning evenvalency metals:
  5. Cd-type [9]) (bivalent, hexagonal),
  6. Ba-type [5]) (bivalent, cubic space centered),
  7. Pb-type [7]) (fourvalent, cubic face centered),
  8. Pt-type [4]) (eightvalent, cubic face centered).

It may be noticed that the sixvalent metals W and Mo fit in with the bivalent Ba-type. Evidently only two electrons are conduction electrons, and the other four d-electrons are heavily bound.

Fig. 1 shows the mutual positions of the different types, each represented by one of the metals belonging to it.

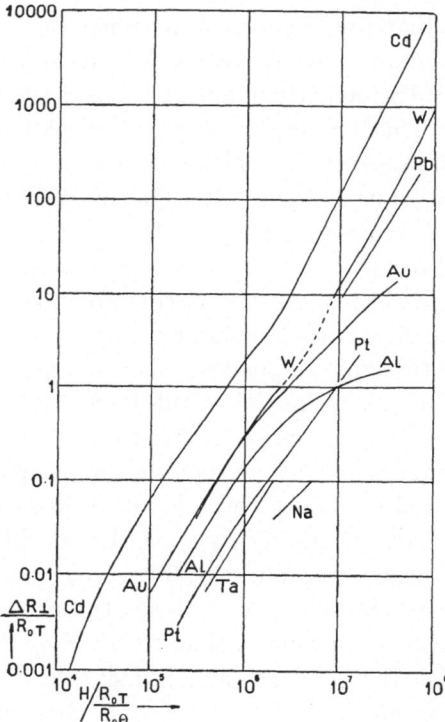

Fig. 1. Reduced Kohler diagram for various metal types.

In general one may say that for the transverse effect for the 1–3–5- and presumably the 7-valent metals there is a small value of $\Delta R/R_{0T}$, whereas for the 2–4–6- and 8-valent metals this value is much higher, especially at the higher values of $H/(R_{0T}/R_{0\Theta})$. For a given value of $H/(R_{0T}/R_{0\Theta})$ the Cd-type shows the highest values of $\Delta R_\perp/R_{0T}$ so far found, with the exception of Bi where the results for different temperatures do not give a single characteristic function in the Kohler diagram.

For the oddvalent metals a saturation tendency exists for the transverse effect, which may have led K a p i t z a [10]) to conclude that the initial square law for the field dependency ends in a subsequent approach to a linear law. His conclusion that this was a general phenomenon shown by all metals has, however, not been confirmed, since the later experiments of J u s t i *c.s.* have revealed that for the evenvalent metals the square law is valid to the highest values of $H/(R_{0T}/R_{0\Theta})$, so far as $\Delta R_\perp/R_{0T}$ is concerned.

Perhaps for $\Delta R_{\parallel}/R_{0T}$ (the parallel effect) the saturation exists also for the evenvalent metals. This has actually been established for the Ba-type metals and M i l n e r's [11]) experiments show it to be highly probable in the case of Cd as well.

4. *The theory of* K o h l e r *on the magneto-resistance effect.* K o h-
l e r [12]) has accounted for his diagram theoretically by extending the
theory of P e i e r l s [13]) on the magneto-resistance effect. P e i-
e r l s was the first to realise that the experimental values of this
effect were very much higher than those derived from either the
classical [14]) or the Sommerfeld [15])[16]) and early quantum theories [17])[18]),
because these theories assumed that the electrons were completely
free. He therefore assumed an anisotropy in the distribution of the
velocities of the electrons but, only by assuming an impossibly
strong anisotropy, could the experimental values of ΔR be realised.

Meanwhile M o t t and J o n e s [19]) developed for the different
crystallographic systems the general form of the "Brillouin zones",
in which the quantummechanically allowed energies of the valency
electrons are classified. Conduction is only possible when the zones
are not totally filled up. When all zones are either full or empty the
material is an insulator at normal temperatures. Now in consequence
of the Fermi statistics two valency electrons are required to fill up a
band in normal circumstances and thus only the oddvalent metals
are good conductors as they have a half filled zone. From this
K o h l e r argues that metals showing relatively small values of
$\Delta R/R_{0T}$ approach the case of free electrons. The tendency to saturation mentioned above for metals with odd valency should according to
the theory also occur in the case of free electrons. The fact that the
parallel effect is not zero, which would be required by the free electron
theory, reveals however that the electrons may only be considered
as comparatively free.

The even valency metals might be presumed to be insulators, but
the "overlapping" of the Brillouin zones, (which M o t t and
J o n e s have established) causes a partial occupation of the second
zone, and such metals will be conductors, although not good conductors. The required electrons are evidently supplied by the first
zone, which consequently is partly unoccupied. The overlapping of
the Brillouin zones must be thought of thus: only in special
directions is the minimum energy in the second zone less than the
maximum energy in the first zone. From this K o h l e r drew attention to the fact that the distribution of the conduction electrons is
strongly anisotropic even for the cubic evenvalent metals.

In addition there are two other causes of anisotropy. Firstly, for
the non-cubic metals the triple periodical field of the nuclei in which

the electrons are moving is different for the three principal directions. Secondly, the temperature dependent perturbations in the lattice will in general also show an anisotropy. This is connected with an anisotropy in the elastic properties, which is greatest in the first column of the periodic system and diminishes gradually from column to column.

By allowing for these effects in the calculation of the magneto-resistance effect, K o h l e r showed that ΔR should only be independent of the orientation of the field with respect to the crystal axes for the 3-, 5- and 7-valent metals when they are cubic. Since these metals are only rarely cubic this implies that the independency occurs only in exceptional cases, *e.g.* for the Al-type where J u s t i *c.s.* have found this property.

For the Na- and Au-types the experimentally determined anisotropy may be principally due to the elastic effect. On the other hand for the Ba-, Pb- and Pt-types it may be caused principally by the overlapping effect.

The most interesting results have been found for the Cd-type where all three reasons for anisotropy exist together.

For the bivalent cubic metals (Ba-type) K o h l e r has, after introducing several simplifications, calculated the value of $\Delta R_{\perp}/R_{0T}$, and he shows that it should follow a square law as a function of $H/(R_{0T}/R_{0°C})$, which remains valid for high values of $H/(R_{0T}/R_{0°C})$, provided that $\mu H \ll kT$ (μ is the magnetic moment of the electron). For $\Delta R_{\parallel}/R_{0T}$ he expects a saturation at high values of $H/(R_{0T}/R_{0°C})$.

The agreement with the experimental data is quite satisfactory and, although the effect appears still too·complicated to permit detailed calculations to be made, K o h l e r's work has made important progress in the understanding of the magneto-resistance effect.

5. *The gallium results in the Kohler diagram. a.* Previously we considered it worth noticing that the fundamental idea of K o h-l e r's theory is identical with a conclusion we drew from our experiments. By the fact that the normal resistance, and its increase in a magnetic field, tend to a temperature independency in the same temperature range [20]), we were led to the view that the increase principally varies with the irregularities in the crystal lattice.

b. Since in our former publications ΔR always has been divided by

$R_{0°C}$ and not by R_{0T} as required for the plotting in the Kohler diagram, it was necessary to divide all the published values by $R_{0T}/R_{0°C}$. In addition the values of H had to be divided by $R_{0T}/R_{0°C}$. We publish here just a few representative values for the different cases.

c. When there is a saturation tendency the division by $R_{0°C}$ led us to an effect which in reality has no physical significance and does not appear when dividing by R_{0T}. We refer to the decreasing of $\Delta R_{\parallel}/R_{0°C}$ which occurs with a constant field when the temperature is lowered, while at a certain temperature a maximum occurs [21].

6. *The transverse effect for the P_1-crystal.* In fig. 2 for both cases when the field is parallel to the short axes (P_2- and P_3-axes) a large number of measuring points have been plotted in the Kohler diagram.

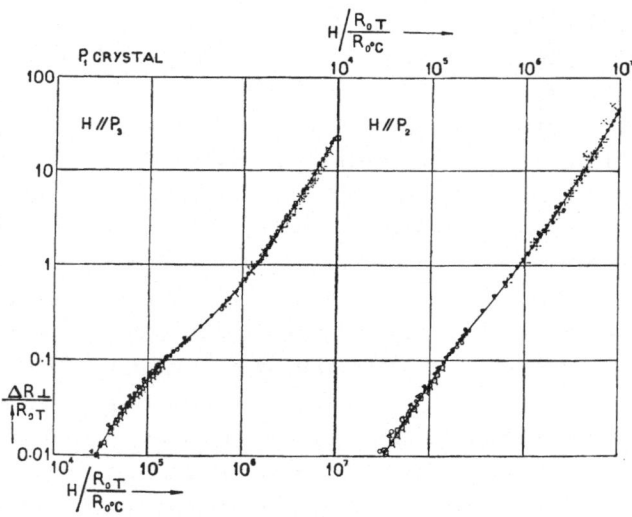

Fig. 2. Kohler diagram of the transverse magneto-resistance effect for a gallium P_1-crystal.

$\odot$ $T = 10.4$ °K	$\boxdot$ $T = 10.4$ °K (ann.)	$\lhd$ $T = 49.8$ °K
$+$ $T = 14.2$ °K	$\times$ $T = 14.2$ °K (ann.)	$\circledcirc$ $T = 56.5$ °K
$\triangle$ $T = 20.4$ °K	$\triangleright$ $T = 20.4$ °K (ann.)	$\boxdot$ $T = 70.2$ °K

a. It is evident from the curves that in both cases a single characteristic function exists representing the relation between $\Delta R_{\perp}/R_{0T}$ and $H/(R_{0T}/R_{0°C})$. The curves found for the different temperatures are seen to agree fairly well, the scatter being greatest for the weak field points where the inaccuracy is rather high, but even here the disagreement is not, in general, more than a few per cent.

The fact that the results measured with different crystals fit in satisfactorily supplies in itself no valuable support for the use of the Kohler diagram, because the different crystals in general show equal values of $R_{0T}/R_{0°C}$, which demonstrates the similarity of the perturbations in the lattice. This, we formerly assumed [22]), might be attributed to the fact that the crystals were grown in the same way from the same material (with the smallest chemical impurities). In consequence the results are of themselves comparable.

However we were able to show that the results for crystals with different lattice irregularities also lie on the same curve. For a crystal, exposed to an annealing process for two years by keeping it a few degrees below its melting point, had its temperature independent perturbations considerably reduced [21]) and now it is found that the results obtained for this crystal fit the curve very well.

b. From the absence of the singularities in the field dependency of $\Delta R_\perp/R_{0°C}$ it might be concluded that gallium behaves more like the normal metals and not like bismuth (the idea with which we initially started our experiments). The fact that the Kohler diagram is valid for gallium and not for bismuth gives a strong support for this conclusion.

The invalidity of the Kohler diagram for bismuth may be inferred from the fact that the singularities in the field dependency of $\Delta R_\perp/R_{0°C}$ occur at the same fields for the different temperatures [23]). As in the Kohler diagram $H/(R_{0T}/R_{0°C})$ is the coordinate and $R_{0T}/R_{0°C}$ is varying with the temperature, the results for the different temperatures can never give rise to coinciding singularities.

c. After the success achieved in general with the Kohler diagram for gallium, a difficulty unfortunately arises concerning the type amongst which this metal should be classified. As it is trivalent and has, like indium, two s-electrons and one p-electron, one would also expect to find the saturation tendency shown for this and the other oddvalent metals. Since, however, we find that this does not occur, for even at the highest values of $H/(R_{0T}/R_{0°C})$ a square law is still valid, it seems much more likely that gallium follows the evenvalent metal types.

d. By using the Kohler diagram a special improvement is found for the study of the interchanging of the shorter axes and this is

6

illustrated in fig. 3, where both the curves, given separately in fig. 2, are combined. In table I representative sets of calculated values are collected for the case of the field parallel to the P_2-axis. In table II the same thing is done for the case of the field parallel to the P_3-axis. The first columns give $H/(R_{0T}/R_{0°C})$ the second columns $\Delta R_\perp/R_{0°C}$

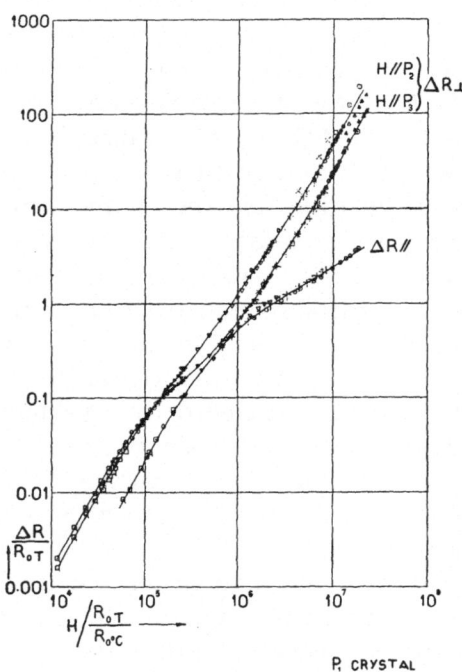

Fig. 3. Kohler diagram of the transverse and the parallel effect for a gallium P_1-crystal.

△ $T =$ 4.2 °K	× $T =$ 14.2 °K (ann.)	◎ $T =$ 56.5 °K
⊙ $T =$ 10.4 °K	▽ $T =$ 20.4 °K	⊡ $T =$ 70.2 °K
⊡ $T =$ 10.4 °K (ann.)	▷ $T =$ 20.4 °K (ann.)	⊡ $T =$ 77.4 °K
⊹ $T =$ 14.2 °K	◁ $T =$ 49.8 °K	

and the third columns the temperature (and where necessary the information that the annealed crystal is concerned).

Apart from the effect that the shorter axes are quite distinguishable (the ratio for both values is about 2) in spite of the pseudo-tetragonal structure of the crystals (which is also discernible without a Kohler diagram), the further details of the difference between these axes are now visible in one diagram. Formerly several field and

TABLE I

TABLE II

$\Delta R_\perp / R_0 T$ for a P_1-crystal with $H//P_2$-axis		
$H\left\lvert\dfrac{R_0 T}{R_0°C}\right\rvert.10^{-6}$	$\Delta R_\perp / R_0 T$	T °K
22.32	161.4	4.22	
18.74	195.4	10.4	annealed
16.53	94.70	4.22	
12.56	73.31	10.4	
10.35	64.82	10.4	annealed
8.48	50.18	14.2	,,
6.78	22.52	14.2	
6.28	20.03	10.4	
4.23	15.46	14.2	annealed
2.65	5.95	20.4	,,
2.48	4.72	14.2	
2.38	4.66	20.4	
1.41	2.26	20.4	annealed
0.267	0.211	49.8	
0.24	0.214	20.4	
0.213	0.157	56.5	
0.140	0.0853	70.2	
0.114	0.0622	77.4	
0.090	0.0546	49.8	
0.062	0.0303	56.5	
0.0273	0.0082	70.2	
0.0117	0.0016	77.4	

$\Delta R_\perp / R_0 T$ for a P_1-crystal with $H//P_3$-axis		
$H\left\lvert\dfrac{R_0 T}{R_0°C}\right\rvert.10^{-6}$	$\Delta R_\perp / R_0 T$	T °K
22.32	109.4	4.22	
17.84	63.30	10.4	annealed
16.69	65.47	4.22	
11.50	31.39	10.4	
10.00	22.78	10.4	annealed
8.23	16.14	14.2	
7.02	13.05	10.4	
6.78	11.06	14.2	
4.18	5.30	14.2	annealed
2.56	2.48	20.4	,,
2.37	2.26	20.4	
2.30	2.01	14.2	
1.35	0.970	20.4	annealed
0.269	0.166	49.8	
0.24	0.164	20.4	
0.214	0.132	56.5	
0.140	0.0850	70.2	
0.114	0.0678	77.4	
0.091	0.0616	49.8	
0.062	0.0346	56.5	
0.0273	0.0086	70.2	
0.0116	0.0020	77.4	

temperature dependencies had to be considered before one could have a total insight into this problem.

e. Considering the curves it will be seen that:

1. For high values of $H/(R_{0T}/R_{0°C})$ the highest value of $\Delta R_\perp / R_{0T}$ is found when the field is parallel to the P_2-axis. This we measured at liquid hydrogen [24]) and liquid helium [20]) temperatures.

2. For low values of $H/(R_{0T}/R_{0°C})$ the highest values of $\Delta R_\perp / R_{0T}$ are found when the field is parallel to the P_3-axis. This we measured at liquid ethylene temperatures [25]).

3. For intermediate values of $H/(R_{0T}/R_{0°C})$ in consequence there is an interchanging of the P_2- and P_3-axes. This we measured at liquid nitrogen temperatures [25]).

4. For $H/(R_{0T}/R_{0°C}) = 1.4 \times 10^5$ the curves intersect. Here we have equivalence of the shorter axes and therefore a full tetragonal symmetry.

The field strength for which this occurs is consistent with the formula:

$$H = 1.4 \times 10^5 \times R_{0T}/R_{0°C}$$

and it is therefore a function of the temperature, since $R_{0T}/R_{0°C}$ is temperature dependent. Actually the experiments showed that this H decreased with temperature.

5. The interchanging of the shorter axes is caused by an anomalous behaviour of gallium when the field is parallel to the P_3-axis, or, which amounts to the same thing, when the field is simultaneously perpendicular to the current and to the P_2-axis. For this case there is a bend in the curve, which in the other curve is practically nonexistent.

6. In the range of the intersection the anomalous curve is gradually changing from the square law to a linear law which led M i l- n e r [11]) after discussing our experiments to the assumption that for this case a saturation tendency exists. The Kohler diagram makes clear that this conclusion is premature, for at higher field strengths the square law comes back again. At liquid nitrogen temperatures this occurs however at field strengths which are beyond those at our disposal. On the other hand at liquid hydrogen temperatures the linear part occurs below the normally used field strengths, where the accuracy of the measurements is very low.

7. *The parallel effect for the P_1-crystal.* The parallel effect has been drawn in the same diagram as the transverse effect (fig. 3). Some calculated values are given in table III, with the same composition as table I.

a. For low values of $H/(R_{0T}/R_{0°C})$ a square law is valid and is found experimentally at liquid nitrogen temperatures; for high values of $H/(R_{0T}/R_{0°C})$ a linear law is reached, while in the intermediate range the power is gradually changing. Consequently at 20.4°K the field dependency curve plotted in a linear scale, may be divided in two parts [21]):

1. a parabolic part,
2. a second part, where ΔR increases more slowly with H, and is

TABLE III

$\Delta R_{||}/R_0 T$ for a P_1-crystal

$H\left/\dfrac{R_0 T}{R_0°C}\right.\cdot 10^{-6}$	$\Delta R_{||}/R_0 T$	T °K
18.74	3.92	10.4
8.38	2.33	14.2
2.53	1.14	20.4
1.51	0.69	10.4
0.67	0.345	14.2
0.245	0.102	49.8
0.20	0.068	20.4
0.196	0.0727	56.5
0.130	0.0365	49.8
0.129	0.0367	70.2
0.109	0.0263	77.4
0.103	0.0250	56.5
0.0674	0.0104	70.2
0.0577	0.0083	77.4

seen from a logarithmic plot to follow a law with power less than unity.

At 14.2 and 10.4°K the parabolic part escapes detection by vanishing beneath the lowest accurately known field strengths.

b. The maximum in the temperature dependency of $\Delta R_{||}/R_{0°C}$ for a certain field strength as a consequence of the division by $R_{0°C}$ has already been noted.

c. A saturation for high values of $H/(R_{0T}/R_{0°C})$ is not completely certain yet but it is quite possible. Probably this is a general effect valid for all metals, evenvalent as well as oddvalent.

d. The ratio $\Delta R_{\perp}/\Delta R_{||}$ is changing with the value of $H/(R_{0T}/R_{0°C})$ in various ways, dependent on which of the two possible cases for $\Delta R_{\perp}$ is considered:

1. When $\Delta R_{\perp}$ is selected with the field parallel to the P_2-axis $\Delta R_{\perp}/\Delta R_{||}$ has for low values of $H/(R_{0T}/R_{0°C})$ the normal value 2, which has been investigated experimentally at liquid nitrogen temperatures. For high values of $H/(R_{0T}/R_{0°C})$ the quotient gradually increases; the maximum really determined is 50 for 10.4°K and 22.30 kG.

2. When $\Delta R_{\perp}$ is selected with the field parallel to the P_3-axis $\Delta R_{\perp}/\Delta R_{||}$ also has initially the value 2. Here we find, however, for increasing values of $H/(R_{0T}/R_{0°C})$ a decrease of $\Delta R_{\perp}/\Delta R_{||}$; the minimum we found is a little above 1, for 20.4°K and low field strengths. For further increasing values of $H/(R_{0T}/R_{0°C})$ an increase of $\Delta R_{\perp}/\Delta R_{||}$ occurs again. The maximum investigated for 10.4°K and 22.30 kG is 17.

The initial approach and the subsequent divergence have however different causes. The approach is due to the anomalous behaviour of the $\Delta R_{\perp}/R_{0T}$-curve, the divergence to the start of the linear part in the $\Delta R_{||}/R_{0T}$-curve.

8. *The transverse effect for the P_2-, P_3- and $P_{2,3}$-crystals.* While in the preceding pages the advantage of the Kohler diagram chiefly lies in the ease with which the results may be surveyed, here we shall see that it also enables us to compare the results from different crystals. It is self-evident that different crystals are used, since the current is parallel to different axes. Inevitably the impurities are unequal and so consequently are the temperature dependencies of $R_{0T}/R_{0°C}$. This has made the conclusions from our previous comparison rather un-

certain. Since now we can use the Kohler diagram, which is independent of these effects, most of these conclusions[22] have been confirmed. It was possible to compare our former results from different crystals because the amount of impurity was reasonably constant, since the crystals were prepared by the same method from the same material.

In the tables IV (P_2-crystal), V (P_3-crystal) and VI ($P_{2,3}$-crystal) some calculated values are assembled. Fig. 4 gives the Kohler diagram.

a. For these crystals there is a greater difference between the values of $\Delta R_\perp/R_{0°C}$ in the extreme cases than for the P_1-crystal. This is plausible as there is a greater difference between the long axis and the shorter ones than between the two short axes themselves.

b. The values of $\Delta R_\perp/R_{0°C}$ for the $P_{2,3}$-crystal are correctly situated between those for the other crystals. Before the results were plotted in this way in the Kohler diagram, small differences in purity and in temperature gave some apparent disagreements.

c. Contrary to expectation the values for the $P_{2,3}$-crystal, when the field is perpendicular to the P_1-axis, do not lie between those for the other crystals at liquid nitrogen temperatures, but are some few per cent lower. Formerly we wrongly attributed this to the varying irregularities in the lattice, but it seems to be a real effect.

d. This may perhaps be connected with another effect found for these crystals, namely when the field is parallel to the P_1-axis the interchanging of the shorter axes is found again, whereas it does not appear when the field is perpendicular to the P_1-axis. Since both anomalies would disappear when for the P_3-crystal, with the field perpendicular to the P_1-axis, all values of $\Delta R_\perp/R_{0°C}$ were somewhat lower, one may suggest that an experimental error has caused too high values of $\Delta R_\perp/R_{0°C}$ and has given rise to these effects.

There are two cases which we considered in connection with those effects. Firstly a deflection of the current from the perpendicular so that it happened to lie in the plane through the P_3-axis and one of the other axes. This however would produce an effect which is just the opposite of that actually observed and therefore cannot be the reason for it. Secondly a deviation of the P_3-axis from the direction of the current so that it happened to lie in the plane through the P_3- and P_1-axes. However the deviation required to account for the observed results is so large that it is quite improbable. It seems

TABLE IV

$H\left|\dfrac{R_0T}{R_0°C}\right|\cdot10^{-6}$	$\Delta R_\perp/R_0T$		T °K
	$H\,/\!/\,P_1$	$H\perp P_1$	
14.97		11.20	10.4
14.39	156.9		10.4
7.99	63.77		14.2
7.92	51.50		10.4
7.91		4.64	14.2
7.81		5.37	10.4
4.24	20.55		14.2
4.18		2.13	14.2
2.88	10.57		20.4
2.82		1.23	20.4
1.52	3.69	0.580	20.4
0.21	0.111	0.0525	56.5
0.117	0.0389	0.0212	77.4
0.113	0.0396	0.0204	56.5
0.0618	0.0132	0.0080	77.4

$\Delta R_\perp/R_0T$ for a P_2-crystal

TABLE V

$H\left|\dfrac{R_0T}{R_0°C}\right|\cdot10^{-6}$	$\Delta R_\perp/R_0T$		T °K
	$H\,/\!/\,P_1$	$H\perp P_1$	
27.53		151.2	10.4
18.12	193.0	69.95	10.4
9.96		25.03	14.2
9.67	63.87		10.4
9.33	60.81		14.2
5.27		9.51	14.2
5.00	22.58		14.2
2.97		3.73	20.4
2.91	9.42		20.4
2.07		2.28	10.4
1.61		1.53	20.4
1.52	3.70		20.4
0.199	0.163	0.0731	56.5
0.113	0.0725	0.0284	77.4
0.105	0.0728	0.0258	56.5
0.0601	0.0259	0.0100	77.4

$\Delta R_\perp/R_0T$ for a P_3-crystal

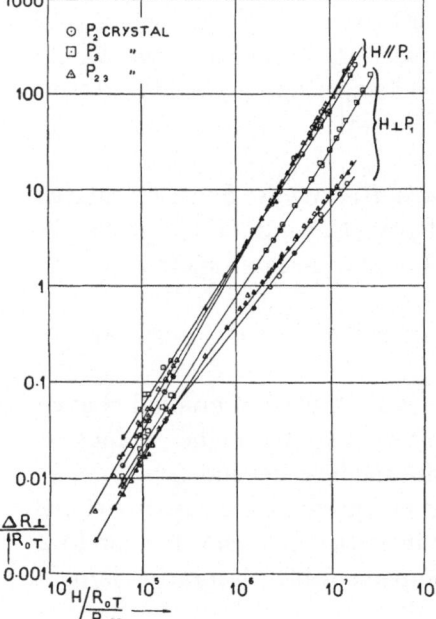

TABLE VI

$H\left|\dfrac{R_0T}{R_0°C}\right|\cdot10^{-6}$	$\Delta R_\perp/R_0T$		T °K
	$H\,/\!/\,P_1$	$H\perp P_1$	
17.80	228.8		10.4
16.89	204.1	18.05	10.4
9.03		8.06	14.2
8.93	77.75		14.2
3.00	11.40	1.83	20.4
2.67	7.30		10.4
2.55		1.60	10.4
1.31	2.83	0.762	14.2
0.464	0.579	0.179	20.4
0.236	0.167	0.0631	56.5
0.129	0.0529	0.0213	77.4
0.057	0.0163	0.0068	56.5
0.0312	0.0044	0.0022	77.4

$\Delta R_\perp/R_0T$ for a $P_{2,3}$-crystal

Fig. 4. Kohler diagram of the transverse effect for P_2-, P_3- and $P_{2,3}$-crystals.

therefore that the anomalies cannot be due to such accidental deviations.

e. Because of the near approach of the curves when the field is parallel to the P_1-axis the point of intersection cannot be determined very precisely. Approximately we derive from the diagram $H/(R_{0T}/R_{0°C}) = 10^6$. Hence the field strength for which the shorter axes are equivalent is consistent with the formula:

$$H = 10^6 \times R_{0T}/R_{0°C}$$

and therefore depends on the temperature.

The difference with the field strength for which the interchanging of the shorter axes occurs for the P_1-crystal is not negligible; for it is greater than the experimental error. (For the P_1-crystal we found $H = 1.4 \times 10^5 \times R_{0T}/R_{0°C}$).

Now the intersection would be moved up considerably by a small variation of the P_3-curve, such as would be connected with the previously considered deviation of the P_3-axis with respect to the current. But, although on the one hand this could improve the agreement with the P_2-value, on the other hand the $P_{2,3}$-curve which now goes satisfactorily through the intersection of the other two, would then be in disagreement. Only renewed experiments would enable us to clear up all uncertainties in this connection.

9. *The parallel effect for the P_2- and P_3-crystals.* Some calculated values are given in the tables VII (P_2-crystal) and VIII (P_3-crystal). In the Kohler diagram (fig. 5) the P_1-curve is again plotted for the purpose of comparison.

a. Naturally the differences between the P_2- and P_3-crystals are less than those with the P_1-crystal.

b. Just as in the case of the P_1-crystal, there is a gradual change from the initial square law to a linear law for higher values of $H/(R_{0T}/R_{0°C})$. Consequently here again the field dependency curve at liquid hydrogen temperatures may be subdivided in two parts, a parabolic, and a second part which increases more slowly. For 14.2 and 10.4° K the parabolic part is imperceptible. Saturation is possible but has not been investigated [26]).

c. As the P_2- and P_3-curves again intersect we find the interchanging of the shorter axes once more. Using the Kohler diagram this conclusion seems quite definitely proved.

TABLE VII

$H\left|\dfrac{R_0T}{R_0°C}\right.\cdot 10^{-6}$	$\Delta R_{||}/R_0T$	T °K
14.87	2.31	10.4
7.96	1.50	14.2
2.84	0.662	20.4
0.811	0.189	10.4
0.307	0.061	20.4
0.216	0.032	14.2
0.211	0.0272	56.5
0.117	0.0092	77.4
0.111	0.0098	56.5
0.0617	0.0031	77.4

$\Delta R_{||}/R_0T$ for a P_2-crystal

TABLE VIII

$H\left|\dfrac{R_0T}{R_0°C}\right.\cdot 10^{-6}$	$\Delta R_{||}/R_0T$	T °K
27.53	1.86	10.4
14.57	1.46	10.4
11.21	2.00	14.2
5.60	1.33	14.2
3.05	0.847	20.4
0.307	0.049	20.4
0.211	0.0162	56.5
0.117	0.0058	77.4
0.114	0.0058	56.5
0.0609	0.0031	77.4

$\Delta R_{||}/R_0T$ for a P_3-crystal

d. On account of the very small difference between the two curves for the P_2- and P_3- crystals the intersection again cannot be determined exactly. Approximately we find $H/(R_{0T}/R_{0°C}) = 10^6$, thus the field strength for which the shorter axes are equivalent is consistent with

$$H = 10^6 \times R_{0T}/R_{0°C}.$$

Again the difference between this and the value found for the P_1- crystal $(1.4 \times 10^5 \times R_{0T}/R_{0°C}$ is greater than the experimental error.

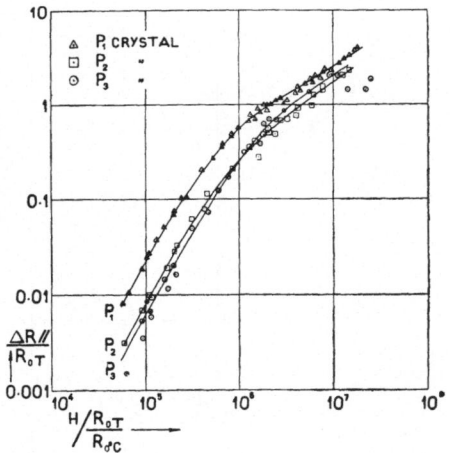

Fig. 5. Kohler diagram of the parallel effect for P_1-, P_2- and P_3-crystals.

10. *Gallium in the reduced Kohler diagram.* In order to be able to investigate the type to which gallium corresponds, our results have been plotted in a reduced Kohler diagram. The difficulty with respect to the Debije temperature has already been mentioned and it proved to be particularly serious for gallium. For on the one hand with the Lindemann formula: $\Theta = 137(T_s/MV^{2/3})^{1/2}$ the result is 125°K, on the other hand specific heat measurements [27] and the temperature dependency of the electrical resistance both give approximately $\Theta = 220$°K (by means of the Grüneisen [28] formula we

estimated for the P_1-crystal the value 210°K, for the P_2-crystal 230°K and for the P_3-crystal 220°K).

Now since J u s t i uses the Lindemann formula for his classification of the metallic types, we should perhaps also use this value 125°K in order to establish to which type gallium belongs. However, we prefer to use the value of 220°K,

a. because this value being based on the Grüneisen formula already has some relations to the electrical conductivity and is more reliable,

b. because we presume that in future all other workers will prefer to use these values.

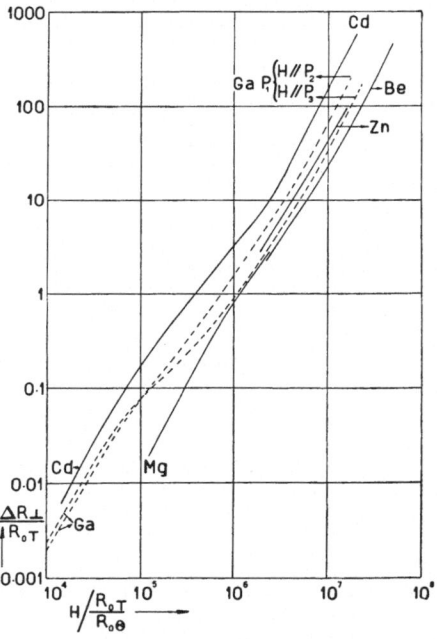

Fig. 6. Reduced Kohler diagram for gallium and the metals of the Cd-type.

Thus, presuming Θ to be 220°K we derived $R_\Theta/R_{0°C}$ with which all values of $H/(R_{0T}/R_{0°C})$ must be multiplied to get $H/(R_{0T}/R_{0\Theta})$ from the temperature dependency of $R_{0T}/R_{0°C}$ and found an average of 0.79 for the three crystals (with $\Theta = 125°K$, $R_\Theta/R_{0°C} = 0.40$ so there is a greater difference than is usually found for the other metals).

Though it would be more logical to use the corresponding Θ for each crystal, we have considered it sufficient to use the average since the values given above are only rough estimations. Like other metals gallium does not have a single Θ-value for all temperature ranges, so we have had to choose values related to the intermediate temperature range.

In fig. 6 the curves for Cd, Zn, Mg and Be, the metals belonging to the Cd-type, based on J u s t i' s data, are drawn together with the gallium results.

It is seen from the curves that the gallium values are less than the cadmium values and greater than the beryllium results. So gallium may be classified in the Cd-type where Cd shows the greatest and Be the smallest values.

Just how well gallium does fit into the Cd-type may be seen from the fact that the axes ratio (1.67) which, as J u s t i has suggested, gives the order of position in the group, lies between that for Cd (1.886) and Be (1.585).

So, for the second time the trivalent gallium seems to belong to the evenvalent metals, although the valency electrons have the same configuration as those of indium, which normally belongs to the odd-valent type (previously we came to this conclusion because of the absence of saturation at high values of $H/(R_{0T}/R_{0°C})$ and now we see there is the correct order of magnitude as well).

However, it will not be possible to give an explanation of the behaviour of gallium or indium until a more complete theory of the Brillouin zones has been developed for these metals, nor can the resemblance of the magneto-resistance effect of gallium to that of the Cd-type be understood before then.

REFERENCES

1) M. K o h l e r, Ann. Physik (5) **32**, 211, 1938.
2) E. J u s t i and H. S c h e f f e r s, Phys. Z. **39**, 105, 1938.
3) E. J u s t i, Phys. Z. **41**, 563, 1940.
4) A. M a t t h i e s s e n and C. V o g t, Ann. Phys. Chem. (Pogg. Folge) **122**,19, 1864.
5) E. J u s t i and J. K r a m e r, Phys. Z. **41**, 197, 1940.
6) E. J u s t i and J. K r a m e r, Phys. Z. **41**, 105, 1940.
7) E. J u s t i, Phys. Z. **41**, 486, 1940.
8) A. F o r o u d, E. J u s t i and J. K r a m e r, Phys. Z. **41**, 113, 1940.
9) E. J u s t i, J. K r a m e r and R e i n h a r t S c h u l z e, Phys. Z. **41**, 308, 1940.
10) P. K a p i t z a, Proc. roy. Soc., London A **123**, 292, 1929.
11) C. J. M i l n e r, Proc. roy. Soc., London A **160**, 207, 1937.
12) M. K o h l e r, Phys. Z. **39**, 9, 1938.
13) R. P e i e r l s, Z. Phys. **53**, 255, 1929; Ann. Physik (5) **10**, 97, 1931.
14) N. H. B o h r, Thesis, København 1911.
15) A. S o m m e r f e l d and N. H. F r a n k, Z. Phys. **47**, 1, 1928.
16) A. S o m m e r f e l d and N. H. F r a n k, Rev. mod. Phys. **3**, 1, 1931.
17) F. B l o c h, Z. Phys. **52**, 555, 1929; **59**, 208, 1930.
18) R. P e i e r l s, Ann. Physik (5) **4**, 123, 1930; (5) **5**, 244, 1930.
19) N. F. M o t t and H. J o n e s, The theory of the properties of metals and alloys (Clarendon Press, Oxford) 1936.
20) W. J. d e H a a s and J. W. B l o m, Commun. Kamerlingh Onnes Lab., Leiden No. 237*d*; Physica, 's-Grav. **2**, 952, 1935.
21) W. J. d e H a a s and J. W. B l o m, Commun. No. 249*d*; Physica, 's-Grav. **4**, 778, 1937.
22) W. J. d e H a a s and J. W. B l o m, Commun. No. 249*c*; Physica, 's-Grav. **4**, 767, 1937.
23) W. J. d e H a a s, J. W. B l o m and L. S c h u b n i k o w, Commun. No. 237*b*; Physica, 's-Grav. **2**, 907, 1935.
24) W. J. d e H a a s and J. W. B l o m, Commun. No. 229*b*; Physica, s'-Grav. **1**, 134, 1933–1934.
25) W. J. d e H a a s and J. W. B l o m, Commun. No. 231*b*; Physica, 's-Grav. **1**, 465, 1933–1934.
26) J. W. B l o m, Commun. No. 281*a*; Physica, 's-Grav. **16**, 144, 1950.
27) K. C l u s i u s and P. H a r t e c k, Z. phys. Chem. **134**, 243, 1928.
28) E. G r ü n e i s e n, Verh. dtsch. phys. Ges. **15**, 186, 1913; **20**, 36, 1918.

THE FOURIER ANALYSIS OF THE ROTATIONAL DIAGRAMS

THE FOURIER COMPONENTS IN THE KOHLER DIAGRAM

Summary

Fourier analysis of the magnetic increase ΔR of the electrical resistance of single-crystals of gallium (with the longer axis parallel to the current) shows that the characteristics of the diagrams giving ΔR as a function of the orientation of the field at liquid nitrogen temperatures are related to the interchanging of the two shorter crystal axes. This is most probably connected with the anomalous behaviour of the magnetic increase of the resistance when the field is perpendicular both to the current and to one of the shorter axes (P_2-axis).

At liquid hydrogen temperatures, for the transverse effect, the term with a period of 60° gains an abnormal importance. When the angle between field and current is varied the saturation tendency in the parallel effect has a great influence in this temperature range.

When plotted in a Kohler diagram characteristic curves are found for all coefficients.

1. *Introduction.* Certain theoretical considerations, which are based on the symmetry of a single-crystal, introduced by V o i g t [1]), enable us to predict some of the coefficients occurring in the Fourier analysis of the function which gives the dependency of a certain quantity on the direction in which it is measured.

We have worked out this analysis for the effect we have studied, *i.e.* the increase of the electrical resistance of gallium (rhombic) in a magnetic field. Firstly we will confine ourselves to the transverse effect, with the field rotating in the plane perpendicular to the current. Then the increase ($\Delta R_\perp$) depends on the orientation of the field in this plane. The results are particularly simple when the current is parallel to one of the three orthogonal axes and the field is accordingly rotating in the plane containing the two other axes.

In general $\Delta R_{\perp}$ depends on the angle α between the field and one of those axes, according to the relation:

$$\Delta R_{\alpha} = C + \sum_{1}^{\infty} a_n \sin n\alpha + \sum_{1}^{\infty} b_n \cos n\alpha. \tag{A}$$

Now, on account of the fact that planes containing two axes are planes of symmetry all coefficients a_n must vanish. Consequently the equation may be written:

$$\Delta R_{\alpha} = C + \sum_{1}^{\infty} b_n \cos n\alpha. \tag{B}$$

Secondly the experimental information that $\Delta R_{\perp}$ is the same when the field is reversed shows that all odd numbered b_n are zero. So we obtain:

$$\Delta R_{\alpha} = C + \sum_{1}^{\infty} b_{2n} \cos 2n\alpha. \tag{C}$$

Since all odd numbered terms vanish it is more convenient to have the even terms consecutively numbered and in addition c_0 is usually written instead of C. Thus we have finally:

$$\Delta R_{\alpha} = c_0 + \sum_{1}^{\infty} c_n \cos 2n\alpha. \tag{D}$$

Experimentally we find as a rule that the higher numbered terms are smaller than the lower ones and frequently the first coefficient is large enough for the other terms to be neglected. Then we have:

$$\Delta R_{\alpha} = c_0 + c_1 \cos 2\alpha. \tag{E}$$

This is, of course, a simple sine-curve, with a period of 180°. It may be noticed that the equation (E) may also be written in the form:

$$\Delta R_{\alpha} = \Delta R_1 \cos^2 \alpha + \Delta R_2 \sin^2 \alpha, \tag{F}$$

where ΔR_1 and ΔR_2 are the values of ΔR for $\alpha = 0°$ and $90°$ respectively.

When changing gradually from the transverse to the parallel effect, by turning the field in a plane containing two axes, (one of which in the direction of the current), again one often finds the simple relation:

$$\Delta R = d_0 + d_1 \cos 2\beta$$

between ΔR and β, the angle between field and current, which in this case may be written:

$$\Delta R = \Delta R_{\parallel} \cos^2 \beta + \Delta R_{\perp} \sin^2 \beta.$$

Since, however, in some particular circumstances our measurements led to rotational diagrams (giving ΔR in dependence on a or β) which were not simply sinusoidal we calculated the Fourier coefficients referring to these diagrams, using a method suggested by R u n g e [2]). The results justified a restriction to $c_0(d_0)$ and the first six terms, the later ones being within the experimental error.

2. *The transverse effect for the P_1-crystal at liquid hydrogen and liquid helium temperatures.* In the first case non-sinusoidal diagrams have been investigated for the transverse effect with P_1-crystals (having the longer axis parallel to the current) at liquid hydrogen [3]) and liquid helium temperatures [4]). In table I the Fourier coefficients calculated for these diagrams have been collected, column 1 giving

TABLE I

T °K	H	c_0	c_1	c_2	c_3	c_4	c_5	c_6
20.4	6.000	0.0051	−0.0015	−0.00008	0.00004	−0.00001	−0.00002	+0.00002
	11.800	0.0123	−0.0041	−0.00021	0.00023	+0.00003		−0.00001
	18.125	0.0234	−0.0087	−0.00048	0.00070	0.00008	+0.00008	
	22.300	0.0313	−0.0120	−0.00072	0.00138	0.00031		−0.00008
14.2	5.175	0.0057	−0.0019	−0.00006	0.00018	0.00007	−0.00006	+0.00001
	7.750	0.0100	−0.0034	−0.00030	0.00028	0.00002	+0.00007	0.00009
	10.250	0.0151	−0.0053	−0.00045	0.00060	0.00016	0.00009	0.00005
	12.600	0.0206	−0.0075	−0.00081	0.00098	0.00031	0.00015	0.00008
	14.525	0.0260	−0.0097	−0.00097	0.00133	0.00027	0.00034	0.00017
	17.125	0.0333	−0.0127	−0.00156	0.00190	0.00067	0.00052	0.00009
	20.075	0.0433	−0.0171	−0.00233	0.00277	0.00112	0.00085	0.00027
	21.650	0.0495	−0.0198	−0.00281	0.00333	0.00123	0.00104	0.00028
10.4	6.000	0.0090	−0.0029	−0.00032	0.00043	0.00009	0.00008	0.00002
	11.800	0.0266	−0.0095	−0.00134	0.00207	0.00070	0.00043	0.00004
	18.125	0.0577	−0.0224	−0.00243	0.00528	0.00106	0.00136	0.00070
9.9	22.300	0.0925	−0.0409	−0.00614	0.00972	0.00389	0.00233	0.00033
1.35	8.800	0.0251	−0.0082	−0.00164	0.00297	0.00117	0.00088	0.00018
	15.275	0.0679	−0.0255	−0.00492	0.00991	0.00419	0.00312	0.00080
	22.100	0.1321	−0.0547	−0.01022	0.02088	0.01085	0.00838	0.00254

Fourier coefficients for gallium P_1-crystals at liquid hydrogen and liquid helium temperatures

the temperature, column 2 the field strength, column 3 c_0 and the further columns c_1 up to c_6.

Apart from the normal result that all coefficients increase with growing field strength and falling temperature, there are two features which we note in particular:

a. in general c_2 is smaller than c_3 in contrast to the normal behaviour,

b. c_3 (which has the opposite sign to c_1) is gaining in importance
compared with c_1 when the temperature is lowered or the field is
increased.

At 20.4°K c_3 is less than 5% of c_0, at 14.2°K about 7%, at 10.4°K
already 10% and at helium temperatures 15%, whereas c_2 is always
about 5% of c_0 and the higher terms are much smaller.

In fact it was possible to describe all the characteristics of the diagrams in this temperature range by the influence of this third Fourier
coefficient which has a period of 60°.

To illustrate this we have constructed a curve with the three following components:

1. a constant c_0,
2. a term $c_1 \cos 2\alpha$ (period 180°),
3. a term $c_3 \cos 6\alpha$ (period 60°),

where the values of c_0, c_1 and c_3 are derived from table I and are consequently the Fourier coefficients found experimentally. We selected
those concerning 9.9°K and 22.30 kG ($c_0=0.09$, $c_1=-0.04$ and $c_3=0.01$).

In fig.1 the individual terms are plotted as a function of α, in fig. 2
the complete curve has been drawn. The characteristics of the exper-

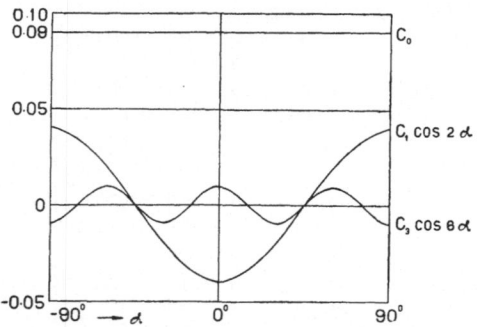

Fig. 1. Separate Fourier coefficients
as a function of α.

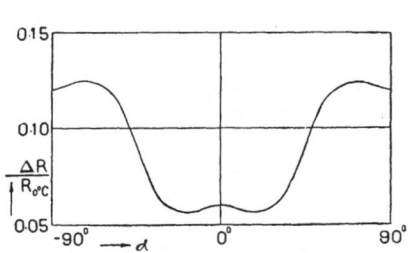

Fig. 2. Constructed rotational
diagram.

imental rotational diagrams [5]) found at the lower temperatures are
seen to be clearly reproduced:

a. The original maximum has been replaced by a minimum, which
is connected with the appearance of two secondary maxima on both
sides of it.

b. The original minimum has been replaced by a maximum, which is connected with the appearance of two secondary minima on both sides of it.

The distances between the secondary maxima and the secondary minima increase with growing field strength and falling temperature since the influence of c_3 increases, compared with that of c_1. In the limit these distances will reach 60°. The experimental data are in good agreement with these conclusions.

3. *The transverse effect for the P_1-crystal at liquid nitrogen and liquid ethylene temperatures.* The second case which we investigated concerned the anomalies in the rotational diagrams for the transverse effect with P_1-crystals which were found at liquid nitrogen temperatures [6]). At liquid ethylene temperatures only a slight effect may still be observed. For these temperatures the Fourier coefficients have also been calculated. They are collected in table II which is arranged in the same way as table I.

TABLE II

T °K	H	c_0	c_1	c_2	c_3	c_4	c_5	c_6
		Fourier coefficients for gallium P_1-crystals at liquid nitrogen and liquid ethylene temperatures						
49.8	6.000	0.00361	+0.00026	—0.00012	0.00001	—0.00001		
	11.800	0.00839	—0.00004	—0.00026	0.00002	—0.00001	0.00001	
	18.125	0.01324	—0.00105	—0.00046	0.00007	—0.00001	0.00001	0.00001
	22.300	0.01630	—0.00187	—0.00058	0.00006	—0.00001	0.00001	0.00001
56.5	6.000	0.00314	+0.00016	—0.00013	0.00001		—0.00001	—0.00001
	11.800	0.00763	+0.00019	—0.00023	0.00004	—0.00002	+0.00001	—0.00001
	15.125	0.01043	—0.00003	—0.00037	0.00003	—0.00003	+0.00003	—0.00001
	18.125	0.01263	—0.00057	—0.00047	0.00007	—0.00002	—0.00001	—0.00001
	22.300	0.01598	—0.00101	—0.00056	0.00007	+0.00002	+0.00002	+0.00001
70.2	11.800	0.00614	+0.00035	—0.00026	0.00004	—0.00004	+0.00001	+0.00001
	18.125	0.01084	+0.00037	—0.00038	0.00004	—0.00005		—0.00001
	24.750	0.01553	—0.00009	—0.00053	0.00007	+0.00002	—0.00001	
77.4	11.800	0.00555	+0.00042	—0.00019	0.00001	—0.00001		—0.00001
	18.125	0.00998	+0.00052	—0.00034	0.00007	—0.00002		—0.00001
	22.300	0.01320	+0.00036	—0.00050	0.00007	—0.00004		—0.00002
126	18.125	0.00595	+0.00060	—0.00015	0.00001	—0.00001	0.00002	0.00001
	22.300	0.00824	+0.00086	—0.00025	0.00003	—0.00002		
155	18.125	0.00449	+0.00049	—0.00007	0.00001	0.00001	0.00002	—0.00001
	22.300	0.00630	+0.00066	—0.00008	0.00003	0.00001		

a. From the results it is evident that the anomalies in the rotational diagrams at liquid nitrogen temperatures are quite different in origin from those at the lower temperatures, since the third term c_3 is negligible here as well as the higher ones (they are invariably less than 0.5% of c_0).

Here it is the first coefficient which shows an unusual behaviour, for we find a change in the sign of c_1 when the temperature or the field strength is varied. At the weaker fields this sign is positive, whereas at the stronger fields it is in general negative.

b. It is possible to describe the particular properties of the rotational diagrams at liquid nitrogen temperatures, as the influence of the interchanging of the shorter axes in this temperature range.

α. The value of c_1 approximately given by:

$$c_1 = \frac{\Delta R_1 - \Delta R_2}{2}$$

(ΔR_1 and ΔR_2 being the values of ΔR when the field is parallel to the P_3- resp. P_2-axis), it is evident that here this term is rather small. In the average c_1 is 5% of c_0, while at liquid hydrogen temperatures we found almost 30%. Accordingly the second term c_2, with a period of 90°, becomes more prominent, and this is demonstrated by the fact that in the diagrams only the original maximum has been displaced. The original minimum which at liquid hydrogen temperatures was also displaced, remains however undisturbed.

β. At liquid ethylene temperatures the shorter axes have completely interchanged compared with the liquid hydrogen temperatures. Here the first Fourier coefficient is again dominant (we found the absolute value of c_1 to be about 10% of c_0), while the influence of the second term (almost 3% of c_0) is imperceptible.

γ. The Fourier coefficients being necessarily continuous functions of the field strength, for each temperature a zero value of c_1 is found for a certain field strength. This field strength decreases with temperature. When c_1 is zero gallium is tetragonal in this respect (we referred to it as pseudo-tetragonal). Since the further coefficients may be neglected this results in a full sinusoidal diagram with a period of 90°. Experimentally we realised this effect under the following conditions: 49.8°K and 11.80 kG, 56.5°K and 15.125 kG, 70.2°K and 24.75 kG.

4. *The dependency on the angle β between the field and the current for the P_1-crystal at liquid nitrogen temperatures.* In a third investigation [7]) second order terms have been found in the rotational diagrams for the P_1-crystal at liquid nitrogen temperatures, when the angle β between field and current is varied (thus changing from the transverse to the parallel effect). Though experimentally this was realised by turning the field round each of the two shorter axes, it was found that a complicated behaviour only occurred when turning round the P_2-axis. Therefore the Fourier coefficients which we called here d_0 up to d_6 have been calculated and collected in table III.

TABLE III

Fourier coefficients for gallium P_1-crystals when the field is turned round the P_2-axis at liquid nitrogen temperatures								
T °K	H	d_0	d_1	d_2	d_3	d_4	d_5	d_6
49.8	11.800	0.0064	—0.0023	—0.0006	—0.0001			
	18.125	0.0105	—0.0020	—0.0012	—0.0004	—0.0001		
	22.300	0.0133	—0.0016	—0.0015	—0.0006	—0.0002	—0.0001	
56.5	11.800	0.0059	—0.0024	—0.0005	—0.0001			
	18.125	0.0099	—0.0026	—0.0011	—0.0003		—0.0001	
	22.300	0.0127	—0.0024	—0.0014	—0.0004	—0.0002	—0.0001	
70.2	11.800	0.0044	—0.0022	—0.0002	—0.0001		—0.0001	
	18.125	0.0082	—0.0032	—0.0007	—0.0001			
	22.300	0.0109	—0.0035	—0.0009	—0.0002			
77.4	11.800	0.0041	—0.0021	—0.0002				
	18.125	0.0077	—0.0031	—0.0007	+0.0001			—0.0001
	22.300	0.0103	—0.0038	—0.0011		+0.0001		—0.0001

a. A certain similarity with the preceding case may be noticed, as here again, at the lower temperatures and the stronger fields, the influence of the first term d_1 is reduced compared with that of d_0. Hence the second term, which at higher temperatures may be neglected, gains in importance. This is represented by the appearance of a secondary minimum which replaces the original maximum, the original minimum remaining unchanged.

b. This decrease of d_1 means here a gradual approach of the parallel effect to the transverse effect for the case of the field parallel to the P_3-axis. It is possible to describe this approach as the influence of the anomalous behaviour of the magnetic increase of the resistance which occurs when the field is perpendicular both to the P_2-axis and to the current. We noted this behaviour when plotting our results in a Kohler diagram[8]). It is evident from the graph that this anomalous transverse effect curve inclines towards that for the parallel effect, when the field is increased or the temperature is lowered.

c. The interchanging of the shorter axes, which causes the modifications in the rotational diagrams for the transverse effect at liquid nitrogen temperatures (see section 3), is also influenced by this anomalous transverse effect, as may be seen from the same Kohler diagram.

So we may finally conclude that at liquid nitrogen temperatures all the modifications (both for the transverse effect and for the β-dependency) are connected with this property of $\varDelta R$ when the field is perpendicular both to the P_2-axis and to the current.

5. *The dependency on the angle β between the field and the current for the P_1-crystal at liquid hydrogen temperatures.* Finally we consider the non-sinusoidal diagrams for the same arrangement as above but at liquid hydrogen temperatures[7]). The Fourier terms are given in table IV.

TABLE IV

T °K	H	d_0	d_1	d_2	d_3	d_4	d_5	d_6
				Fourier coefficients for gallium P_1-crystals when the field is turned round the P_2-axis at liquid hydrogen temperatures				
20.4	6.000	0.0040	+0.0001	—0.0004	—0.0002	—0.0001	—0.0001	
	11.800	0.0082	—0.0003	—0.0006	—0.0005		—0.0002	
	18.125	0.0134	—0.0025	—0.0009	—0.0007	+0.0001	—0.0003	
	22.300	0.0162	—0.0043	—0.0009	—0.0010	0.0002	—0.0005	0.0002
14.2	2.400	0.0017		—0.0002	—0.0001	—0.0001		
	3.600	0.0024	—0.0002	—0.0002	—0.0001			
	4.800	0.0034	—0.0006	—0.0003	—0.0002		—0.0001	
	6.000	0.0044	—0.0010	—0.0003	—0.0002		—0.0001	
	11.800	0.0091	—0.0046	0.0001	—0.0004	+0.0002	—0.0004	0.0002
	18.125	0.0157	—0.0104	0.0011	—0.0012	0.0006	—0.0011	0.0006
	22.300	0.0203	—0.0155	0.0027	—0.0020	0.0009	—0.0013	0.0010
10.4	2.400	0.0016	—0.0004	—0.0001	—0.0001			
	6.000	0.0045	—0.0023		—0.0002		—0.0002	
	11.800	0.0120	—0.0100	0.0017	—0.0011	0.0006	—0.0009	0.0005
	18.125	0.0219	—0.0200	0.0043	—0.0028	0.0014	—0.0024	0.0014
	22.300	0.0312	—0.0298	0.0068	—0.0045	0.0022	—0.0028	0.0017

a. Here the importance of the first term compared with d_0 is increasing again with falling temperature and increasing field. Accordingly there is a disappearance of the secondary minimum which replaces the original maximum. Since neither d_2 nor the next terms may be neglected here, the full sine-curve fails to reappear (it may be noticed that d_4 is less than d_5 which is quite unusual).

Only at the low field strengths do we occasionally find small values of d_1 (corresponding with a secondary minimum in the diagram), thus proving the continuous course of the effect, compared with the liquid nitrogen temperatures. At 14.2°K and 2.40 kG a zero value is found. The diagram is, however, not a simple sine-curve with a period of 90°, owing to the influence of d_3 and further terms.

b. Just as with the transverse effect in this temperature range, the complications in the diagrams must have quite another origin than those at the liquid nitrogen temperatures. The first term d_1 being approximately $(\Delta R_{||} -\!\!- \Delta R_{\perp})/2$, the reappearance of the original maximum must be seen as the influence of the subsequent divergence of the transverse and the parallel effect after the initial approach, which may be noted in the Kohler diagram mentioned before.

This divergence is obviously connected with the anomalous course of the parallel effect at the liquid hydrogen temperatures. Although it could not be investigated, as the maximum field available was not sufficient, a saturation seems quite possible there.

c. Here it is the second coefficient which shows a change in the sign. At the weaker fields it is negative, in accordance with the liquid nitrogen results, whereas at the stronger fields it is positive.

6. *The Fourier coefficients in the Kohler diagram.* The use of the Kohler diagram [9]) is very successful for referring to the values of ΔR in the extreme positions [8]), so, assuming it is valid for arbitrary field orientations, we decided to apply a similar treatment to the Fourier coefficients. We again expected to find characteristic curves independent of the temperature or of the purity of the material.

For the P_1-crystal the values of $c_0/(R_{0T}/R_{0°C})$ and $c_1/(R_{0T}/R_{0°C})$ up to $c_6/(R_{0T}/R_{0°C})$ have been calculated for the transverse effect. The results are given in table V, the first column giving the temperature, the second $H/(R_{0T}/R_{0°C})$ and further columns $c_0/(R_{0T}/R_{0°C})$ up to $c_6/(R_{0T}/R_{0°C})$. In fig. 3 the absolute values of the different terms have been plotted against $H/(R_{0T}/R_{0°C})$.

a. In accordance with the expectations all points lie, within the

TABLE V

T °K	$H/\dfrac{R_0T}{R_0°C}\cdot10^{-6}$	$c_0/\dfrac{R_0T}{R_0°C}$	$c_1/\dfrac{R_0T}{R_0°C}$	$c_2/\dfrac{R_0T}{R_0°C}$	$c_3/\dfrac{R_0T}{R_0°C}$	$c_4/\dfrac{R_0T}{R_0°C}$	$c_5/\dfrac{R_0T}{R_0°C}$	$c_6/\dfrac{R_0T}{R_0°C}$
1.35	22.10	132.1	—54.7	—10.2	20.9	10.8	8.4	2.5
1.35	15.28	67.9	—25.5	— 4.9	9.9	4.2	3.1	0.8
9.9	11.49	47.68	—21.08	— 3.16	5.01	2.00	1.20	0.17
1.35	8.80	25.1	— 8.2	— 1.6	3.0	1.2	0.9	0.2
10.4	8.63	27.48	—10.67	— 1.16	2.51	0.50	0.65	0.33
10.4	5.62	12.67	— 4.52	— 0.64	0.99	0.33	0.20	0.02
14.2	5.27	12.04	— 4.82	— 0.68	0.81	0.30	0.25	0.07
14.2	4.88	10.54	— 4.16	— 0.57	0.67	0.27	0.21	0.07
14.2	4.17	8.11	— 3.09	— 0.38	0.46	0.16	0.13	0.02
14.2	3.53	6.33	— 2.36	— 0.24	0.32	0.07	0.08	0.04
14.2	3.07	5.01	— 1.82	— 0.20	0.24	0.07	0.04	0.02
10.4	2.86	4.29	— 1.38	— 0.15	0.20	0.04	0.04	0.01
14.2	2.49	3.67	— 1.29	— 0.11	0.15	0.04	0.02	0.01
20.4	2.25	3.16	— 1.21	— 0.07	0.14	0.03		—0.01
14.2	1.89	2.43	— 0.83	— 0.07	0.07		0.02	+0.02
20.4	1.83	2.36	— 0.88	— 0.05	0.07	0.01	0.01	
20.4	1.19	1.24	— 0.41	— 0.02	0.02			
20.4	0.61	0.52	— 0.15	— 0.01				
49.8	0.241	0.176	— 0.020	— 0.006	0.001			
49.8	0.196	0.143	— 0.011	— 0.005	0.001			
56.5	0.191	0.137	— 0.009	— 0.005	0.001			
56.5	0.157	0.109	— 0.005	— 0.004	0.001			
70.2	0.140	0.088	— 0.001	— 0.003				
56.5	0.130	0.090		— 0.003				
49.8	0.128	0.091		— 0.003				
77.4	0.108	0.064	+0.002	— 0.002				
56.5	0.103	0.066	0.002	— 0.002				
70.2	0.102	0.061	0.002	— 0.002				
77.4	0.088	0.049	0.002	— 0.002				
70.2	0.067	0.035	0.002	— 0.001				
77.4	0.057	0.027	0.002	— 0.001				
126	0.055	0.0200	0.0021	— 0.0006				
126	0.044	0.0145	0.0015	— 0.0004				
155	0.041	0.0116	0.0012	— 0.0003				
155	0.033	0.0082	0.0009	— 0.0001				

Fourier coefficients for the Kohler diagram of the transverse effect for P_1-crystals

experimental error, on characteristic curves, independent of the temperature and the impurity of the crystal.

    b. With regard to the individual coefficients the following may be remarked.

    α. The average value c_0 has the normal course with a square law valid up to the highest values of $H(R_{0T}/R_{0°C})$.

    β. In consequence of the sign change of the first term c_1, in the logarithmic scale some $+c_1$ values and some $-c_1$ have been plotted (at $H/(R_{0T}/R_{0°C}) = 1.4 \times 10^5$, $\log c_1/(R_{0T}/R_{0°C}) = -\infty$).

γ. The second term c_2 being negative, we were obliged to plot only the magnitude. Normally this term may be neglected compared with c_1. Only in the region $H/(R_{0T}/R_{0°C}) = 1.4 \times 10^5$, where c_1 vanishes, does this coefficient have some influence as a second order effect. This was observed at liquid nitrogen temperatures.

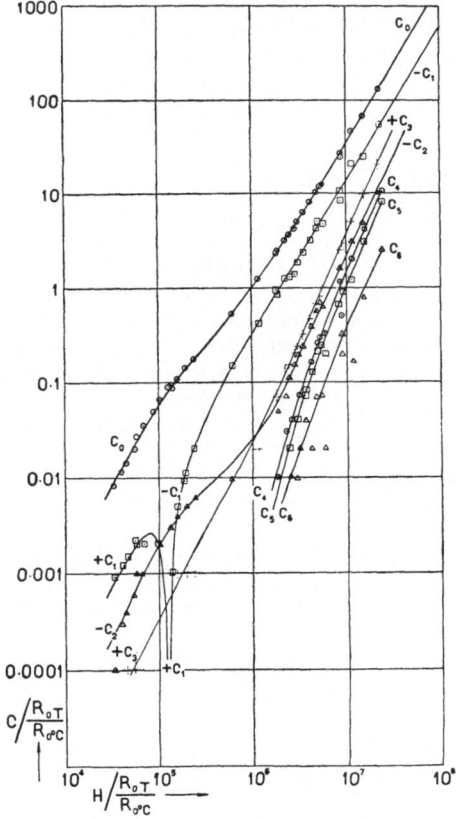

Fig. 3. Fourier coefficients in a Kohler diagram.

δ. The third term c_3, which is negligible at liquid nitrogen temperatures where $H/(R_{0T}/R_{0°C})$ is small, at liquid hydrogen temperatures becomes greater than c_2. This gives rise to the special complications in the rotational diagrams previously investigated both at these temperatures and at liquid helium temperatures.

ε. The further terms c_4, c_5 and c_6 are in general of no importance compared with the preceding terms. Therefore it is a remarkable fact that nevertheless the points are lying satisfactorily on characteristic curves.

c. Now that these seven curves are known, it is possible to predict the increase of the resistance for an arbitrary P_1-crystal (having any impurity) at any arbitrary values of temperature, field strength, and direction of the field in the plane perpendicular to the current, by simply measuring the value of $R_{0T}/R_{0°C}$ (the resistance in zero field), at the temperature chosen.

REFERENCES

1) W. V o i g t, Lehrbuch der Kristallphysik (Teubner, Leipzig) 1910.
2) C. R u n g e, Theorie und Praxis der Reihen (Göschen, Leipzig) 1904.
3) W. J. d e H a a s and J. W. B l o m, Commun. Kamerlingh Onnes Lab., Leiden No. 229*b*; Physica, 's-Grav. **1**, 134, 1933–1934.
4) W. J. d e H a a s and J. W. B l o m, Commun. No. 237*d*; Physica, 's-Grav. **2**, 952, 1935.
5) Cf. reference 3), in particular fig. 6.
6) W. J. d e H a a s and J. W. B l o m, Commun. No. 231*b*; Physica, 's-Grav. **1**, 465, 1933–1934.
7) W. J. d e H a a s and J. W. B l o m, Commun. No. 249*d*; Physica, 's-Grav. **4**, 778, 1937.
8) J. W. B l o m, Commun. Suppl. No. 102*a*; Physica, 's-Grav. **16**, 152, 1950; in particular fig. 3.
9) M. K o h l e r, Ann. Physik (5) **32**, 211, 1938.

SAMENVATTING

De electrische weerstand van gallium éénkristallen ondergaat in een magnetisch veld een toename ΔR, die in sterke mate afhangt van de oriëntatie van stroomrichting en veld ten opzichte van het kristalrooster.

Voor kristallen, waarbij de stroom evenwijdig loopt aan de langste as (P_1-kristallen), bestaat bij het transversale effect in het algemeen een vrij groot verschil tussen de twee gevallen, waarin het veld evenwijdig is aan de ene of de andere der twee korte assen (P_2- en P_3-assen). Deze assen zijn met behulp van de uitwendige vorm in het geheel niet en met behulp van Röntgen-analyse of electrische weerstandsmetingen nauwelijks te onderscheiden, omdat ze gelijke lengte hebben en slechts door een enigszins andere ligging der atomen verschillen.

In het temperatuurgebied van vloeibare stikstof blijken de korte assen bij deze metingen van rol te verwisselen. De temperatuur, waarbij de twee assen gelijkwaardig zijn, hangt af van de veldsterkte; bij lagere veldsterkte ligt deze temperatuur lager.

Deze verwisseling hangt samen met een abnormaal verloop van de temperatuur- en veldafhankelijkheden van ΔR, wanneer het veld loodrecht zowel op de stroom als op de P_2-as staat.

Bij de metingen van het transversale effect voor kristallen met de korte assen evenwijdig aan de stroom (P_2- en P_3-kristallen) speelt deze verwisseling ook een rol. De gelijkwaardigheid wordt hier bij iets lagere temperaturen geconstateerd, wat waarschijnlijk een gevolg is van kleine verschillen in zuiverheid van de gebruikte kristallen.

Daar zowel grotere zuiverheid als lagere temperatuur gepaard gaan met een verhoging van ΔR, lijkt de grootte van de roosterstoringen een maatstaf voor ΔR op te leveren. Hiermede in overeenstemming vindt men, dat in het temperatuurgebied van vloeibaar helium ΔR in gelijke mate temperatuur-onafhankelijk wordt als de gewone weerstand.

Langdurig temperen van de kristallen bij kamertemperatuur blijkt de roosterstoringen aanzienlijk te verminderen.

Wanneer het veld evenwijdig is aan de stroom vertoont ΔR als functie van de veldsterkte bij alle kristallen (P_1-, P_2- zowel als P_3-kristallen) voor hoge velden bij de temperaturen van vloeibare waterstof neiging tot verzadiging.

Als men de resultaten van de metingen uitzet in een Kohler diagram vindt men, zowel voor het transversale als voor het parallelle effect, een karakteristieke functie, die slechts afhangt van de oriëntatie van het veld ten opzichte van de kristalassen en de stroomrichting, maar onafhankelijk is van de temperatuur en de zuiverheid van het gebruikte kristal. Het gebruik van het diagram maakt het mogelijk de resultaten van verschillende kristallen te vergelijken en vergroot de overzichtelijkheid van de metingen, die bij verschillende temperaturen uitgevoerd zijn.

Het blijkt mogelijk gallium in te delen bij één van de door J u s t i ingevoerde metaal-typen. Merkwaardigerwijze is dit echter het 2-waardige Cd-type en niet het type der 3-waardige metalen, waartoe onder meer indium, dat soortgelijke geleidingselectronen zou moeten hebben als gallium, behoort.

Bij Fourier analyse van de draaidiagrammen, die ΔR voorstellen als functie van de richting van het veld, blijken in sommige gevallen termen van hogere orde op de voorgrond te treden. Deze zijn, voor zover het de resultaten bij de temperaturen van vloeibare stikstof betreft, te beschouwen als een gevolg van de rolverwisseling der korte assen.

Wanneer men het veld draait ten opzichte van de stroomrichting is, bij de lagere veldsterkten in het vloeibare waterstofgebied, de bij het parallelle effect optredende neiging tot verzadiging van grote invloed op het gedrag van de Fourier componenten.

Bij het transversale effect wordt bij daling der temperatuur in het vloeibare waterstofgebied langzamerhand een term met een periode van 60° steeds belangrijker.

Wanneer de Fourier coëfficiënten in een Kohler diagram worden uitgezet, vindt men eveneens karakteristieke krommen.

Het afsluiten van dit proefschrift biedt mij een welkome gelegen-
heid het wetenschappelijk en technisch personeel van het Kamerlingh Onnes laboratorium hartelijk dank te zeggen voor de toege-
wijde hulp, waarop ik altijd heb kunnen rekenen bij de voorbereiding
en de uitvoering van de experimenten, die ten grondslag liggen aan
dit proefschrift. Mrs. D a n i e l s ben ik zeer dankbaar voor haar
onmisbare steun bij het vertalen van de tekst.